EXPERIMENTAL
MATHEMATICS
WITH
MAPLE

CHAPMAN & HALL/CRC MATHEMATICS

OTHER CHAPMAN & HALL/CRC MATHEMATICS TEXTS:

**Functions of Two Variables,
Second edition**
S. Dineen

Network Optimization
V. K. Balakrishnan

**Sets, Functions and Logic:
A foundation course in mathematics**
Second edition
K. Devlin

**Algebraic Numbers and Algebraic
Functions**
P. M. Cohn

**Dynamical Systems:
Differential equations, maps and
chaotic behaviour**
D. K. Arrowsmith and C. M. Place

Control and Optimization
B. D. Craven

Elements of Linear Algebra
P. M. Cohn

Error-Correcting Codes
D. J. Bayliss

**Introduction to Calculus of
Variations**
U-Brechtken-Mandershneid

Integration Theory
W. Filter and K. Weber

Algebraic Combinatorics
C. D. Godsil

**An Introduction to Abstract
Analysis-PB**
W. A. Light

The Dynamic Cosmos
M. Madsen

Algorithms for Approximation II
J. C. Mason and M. G. Cox

Introduction to Combinatorics
A. Slomson

Galois Theory
I. N. Stewart

**Elements of Algebraic Coding
Theory**
L. R. Vermani

**Linear Algebra:
A geometric approach**
E. Sernesi

**A Concise Introduction to
Pure Mathematics**
M. W. Liebeck

Geometry of Curves
J. W. Rutter

*Full information on the complete range of Chapman & Hall/CRC Mathematics books is
available from the publishers.*

EXPERIMENTAL MATHEMATICS WITH MAPLE

FRANCO VIVALDI

CHAPMAN & HALL/CRC

Boca Raton   London   New York   Washington, D.C.

Library of Congress Cataloging-in-Publication Data

Vivaldi, Franco
  Experimental mathematics with Maple / Franco Vivaldi.
      p. cm. — (Chapman & Hall/CRC mathematics)
    Includes bibliographical references and index.
    ISBN 1-58488-233-6 (alk. paper)
    1. Mathematics—Data processing. 2. Maple (Computer file) I. Title.
  II. Series.
    QA76.95 .V58 2001
    510—dc21
                                                                    00-052372

Visit the CRC Press Web site at www.crcpress.com

© 2001 by Chapman & Hall/CRC

No claim to original U.S. Government works
International Standard Book Number 1-58488-233-6
Library of Congress Card Number 00-052372
Printed in the United States of America   2 3 4 5 6 7 8 9 0
Printed on acid-free paper

A Chiara e Giulia,
per l'allegria.

Contents

1   What is Maple?   **1**

2   Integers and rationals   **5**
   2.1   Integers   5
   2.2   Arithmetical expressions   8
   2.3   Some Maple   12
   2.4   Divisibility   20
   2.5   Rationals   25
   2.6   Primes   32
   2.7   Standard library functions   37

3   Sets and functions   **39**
   3.1   Sets   39
   3.2   Sets with Maple   45
   3.3   Functions   48
   3.4   User-defined functions   50

4   Sequences   **59**
   4.1   Basics   59
   4.2   Sequences with Maple   61
   4.3   Plotting the elements of a sequence   64
   4.4   Periodic and eventually periodic sequences   67
   4.5   Some non-periodic sequences   70
   4.6   Basic counting sequences   77
   4.7   Sequences defined recursively   82

5   Real and complex numbers   **89**
   5.1   Digits of rationals   89
   5.2   Real numbers   96
   5.3   Random and pseudo-random digits*   99
   5.4   Complex numbers   102
   5.5   Standard library functions   108

6 Structure of expressions **113**
  6.1 Analysis of an expression 113
  6.2 More on substitutions 117
  6.3 Functions acting on operands of expressions 119

7 Polynomials and rational functions **127**
  7.1 Polynomials 127
  7.2 Polynomial arithmetic 128
  7.3 Rational functions 138
  7.4 Basic manipulations 139
  7.5 Partial fractions decomposition* 142

8 Finite sums and products **147**
  8.1 Basics 147
  8.2 Sums and products with Maple 149
  8.3 Symbolic evaluation of sums and products 152
  8.4 Double sums and products 154
  8.5 Sums and products as recursive sequences 160

9 Elements of programming **163**
  9.1 Iteration 163
  9.2 Study of an eventually periodic sequence 173
  9.3 Conditional execution* 177
  9.4 Procedures* 179

10 Vector spaces **185**
  10.1 Cartesian product of sets 185
  10.2 Vector spaces 186
  10.3 Vectors with Maple 188
  10.4 Matrices 191
  10.5 Matrices with Maple 192

11 Modular arithmetic* **201**
  11.1 A modular system 201
  11.2 Arithmetic of equivalence classes 204
  11.3 Some arithmetical constructions in $\mathbb{F}_p$ 206

12 Some abstract structures* **213**
  12.1 The axioms of arithmetic 213
  12.2 Metric spaces 214
  12.3 Rings and fields 214
  12.4 Vector spaces 215

Preface

This text contains sufficient material for a one-semester course in experimental mathematics for first-year students in mathematics or computer science. The course — based on the computer algebra system Maple[1] — introduces the foundations of discrete mathematics while developing basic computational skills. No previous knowledge of computing is assumed.

Contrary to common folklore, mathematical knowledge is seldom acquired through a sequence of flawless logical steps within an abstract framework. A mathematical discovery is more likely to originate from the patiently collected experience of many specific computations. With the aid of computers, contemporary mathematics is beginning to recover the cultural unity that had characterized its tradition until the beginning of the 19th-century, when the boundaries between the pure and the applied side of the discipline were blurred, and when even the greatest minds felt little obligation to hide their calculations and experiments behind a polished axiomatic presentation.

It is in this spirit that we intend to present mathematics to beginning undergraduates. Students often arrive at university without a background in abstract reasoning, and with only a limited experience in the mechanical application of techniques. Developing understanding and familiarity with basic abstract concepts, notation, and jargon (sets, functions, equivalence classes, etc.) is already a considerable task. In order to acquire a taste for abstraction, the students first need to gather *evidence* of mathematical phenomena.

Following an emerging trend in research, we shall place abstraction and axiomatization at the end of a learning process that begins with computer experimentation. Our strategy is to expose the students to a large number of concrete computational examples, repeatedly encouraging them to isolate the general from the particular, to explore, to make a synthesis of computational results, to formulate conjectures, and to attempt rigorous proofs whenever possible.

The choice of an algebraic manipulator — as opposed to a compiled

---

[1] [TM] Maple is a registered trademark of Waterloo Maple, Inc. 57 Erb St. W., Waterloo, Canada N2L 6C2, `www.maplesoft.com`.

language — is central to this strategy, as it allows a straightforward representation of many mathematical objects (such as numbers, sets, polynomials, functions, etc.), thereby eliminating layers of technicalities that, with a compiled language, would invariably lie between the user and the object being represented.

Throughout the course, arithmetic takes centre stage. Arithmetical concepts are transparent and tangible, and possess an appeal to fundamentals seldom found in undergraduate mathematics. Experimentation in arithmetic can be at the same time accessible, illuminating, and rewarding, and Maple's remarkable possibilities will not fail to impress the student.

The arithmetic of integers and rationals is developed in some detail. The real numbers are introduced constructively, from the analysis of the digits of the rationals. Other examples of arithmetical environments include polynomials, rational functions, vector spaces, and modular integers. Some of this material may be treated as optional. (An introductory text, see Reference [3], could provide complementary material and support on the theoretical aspects of this course.)

A second unifying theme is that of a sequence. Sequences are first introduced as explicit functions of the integers. For their Maple representation, we rely on the arrow operator construct -> (functions defined by a single statement), which closely mimics the mathematical notation. Recursive sequences appear later, and are characterized from the perspective of the theory of dynamical systems. They provide examples of complicated phenomena from simple ingredients — an ideal environment for experimentation.

The elements of programming are introduced gradually. Iteration (the do-structure) is delayed until the second part of the course, to allow the concept and notation of a sequence to consolidate. Conditional execution (the if-structure) is introduced and used extensively to define characteristic functions, thereby providing the necessary exposure to logical calculus. Its general form and usage are discussed only towards the end of the course, as optional material. The definition of procedures is also included as optional material. In our view, procedures could be omitted altogether from a first course in computing. In their place, we make extensive use of the arrow operator (one-statement procedures), as well as of structural functions such as seq, map, and select.

Few theorems are stated, often without proof but always exemplified by computations. Several examples of abstract structures (metric spaces, rings, fields, etc.) are given. The corresponding axiomatic definitions are collected together in the last chapter, as a summary of the theory underpinning the course.

Computer work is essential. Many problems are provided, with the more difficult ones flagged by a star. The student is expected to attempt at least half of them. The un-starred problems, particularly those in the later chapters, are meant to have a degree of difficulty comparable to that of examination questions.

Starred chapters, sections, examples, etc., contain optional material, which may require support from other courses, particularly linear algebra.

In this course, Maple is viewed more as a tool than as an object of study in itself. Nevertheless, the material presented here constitutes a fairly exhaustive introduction to the language.

This material has grown from a beginning undergraduate course I have taught at the University of London for five years. I would like to thank all the students who participated in the *'Spot the error'* competition, providing minute inspections of early versions of the manuscript in the hope of winning tickets for some hot West End show. I am also grateful to Sheryl Koral, Leonard Soicher, and Francis Wright, who carefully read various parts of the manuscript at various stages and offered invaluable comments and suggestions. It is now up to the reader to spot the remaining errors (see the book's web page, below).

London   *Franco Vivaldi*

Supplementary material, comments, and errata
can be found on the book's web page:

`http://www.maths.qmw.ac.uk/~fv/emm/`

Chapter 1

What is Maple?

Maple is a powerful system for symbolic computation. In this chapter, we intend to give you a flavour of Maple's capabilities by taking you through a guided tour of some Maple commands. You need a computer with a working copy of Maple software.

Starting and ending a Maple session

The procedure for starting a Maple session depends on the computer and on the details of the software installation. For instance, if you are using a window-based version of the software, you will click on Maple's icon with the mouse. After starting a Maple session, the system will greet you with a *prompt*, which may look like this

>

The *cursor* — the symbol that indicates the place where any typing will end up (typically a bar '|' or an underscore '_') — will be positioned just after the cursor, waiting for input. Whenever Maple displays a prompt, it means that it is ready to accept a new command. One of these commands is `quit` (or `done`, or `stop`), which terminates a Maple session.

> quit

You first type it, then press the RETURN key. In a window environment, you may achieve the same result by clicking on the 'quit' button with the mouse. Before falling silent, Maple will display a brief summary of the resources used during the session, which need not concern you now.

A guided tour of Maple

Here are a few things Maple can do.

Let n be equal to 40! (read '40 factorial'), which is the product of the first 40 positive integers: $40! = 1 \cdot 2 \cdot 3 \cdots 39 \cdot 40$.

> n:=40!;

$$n := 815915283247897734345611269596115894272000000000$$

1

Decompose n into prime factors.

```
> ifactor(n);
```

$$(2)^{38}\,(3)^{18}\,(5)^9\,(7)^5\,(11)^3\,(13)^3\,(17)^2\,(19)^2\,(23)^2\,(29)\,(31)\,(37)$$

(Did you expect 40! to be so large? Does its prime factorization make sense to you? Think about it.)

Display the first 100 digits of π. (Note that `Pi` has upper-case P and lower-case i.)

```
> evalf(Pi,100);
```

$$3.141592653589793238462643383279502884197169399375105\backslash$$
$$82097494459230781640628620899862803482534211706\overline{8}$$

Expand the binomial $(x + \sqrt{y})^{10}$.

```
> expand((x+sqrt(y))^10);
```

$$x^{10} + 10\,x^9\,\sqrt{y} + 45\,x^8 y + 120\,x^7 y^{3/2} + 210\,x^6 y^2 + 252\,x^5 y^{5/2}$$
$$+210\,x^4 y^3 + 120\,x^3 y^{7/2} + 45\,x^2 y^4 + 10\,xy^{9/2} + y^5$$

Compute the finite sum

$$\sum_{k=1}^{n-1} k^3\,5^k.$$

```
> sum(k^3*5^k,k=1..n-1);
```

$$\frac{45}{32}5^n\,n - \frac{115}{128}5^n - \frac{15}{16}5^n n^2 + \frac{1}{4}n^3\,5^n + \frac{115}{128}$$

Compute the limit

$$\lim_{x\to\infty} \frac{3x - 1}{x\arctan(x) + \ln(x)}.$$

```
> limit((3*x-1)/(x*arctan(x)+ln(x)),x=infinity);
```

$$6\,\frac{1}{\pi}$$

Compute the derivative of $\sin\left(x^2\ln(1 - x^3)\right)$.

```
> diff(sin(x^2*ln(1-x^3)),x);
```

$$\cos(x^2\ln(1 - x^3))\left(2\,x\ln(1 - x^3) - 3\,\frac{x^4}{1 - x^3}\right)$$

Compute the indefinite integral

$$\int x^2\sqrt{x^2 - a^2}\,dx.$$

```
> int(x^2*sqrt(x^2-a^2), x);
```
$$\frac{1}{4}x\left(x^2 - a^2\right)^{3/2} + \frac{1}{8}a^2x\sqrt{x^2 - a^2} - \frac{1}{8}a^4 \ln(x + \sqrt{x^2 - a^2})$$

Solve the system of linear equations

$$5x - 3y = 2z + 1, \qquad -x + 4y = 7z, \qquad 3x + 5y = z.$$

```
> eqns:={5*x-3*y=2*z+1,-x+4*y=7*z,3*x+5*y=z};
```
$$eqns := \{5x - 3y = 2z + 1, -x + 4y = 7z, 3x + 5y = z\}$$

```
> solve(eqns,{x,y,z});
```
$$\left\{x = \frac{31}{255}, y = -\frac{22}{255}, z = -\frac{1}{15}\right\}$$

Plot the graph of the function

$$\sin(x^3)\ln(1 + x^2), \qquad x \in [-2, 2].$$

```
> plot(sin(x^3)*log(1+x^2), x=-2..2);
```

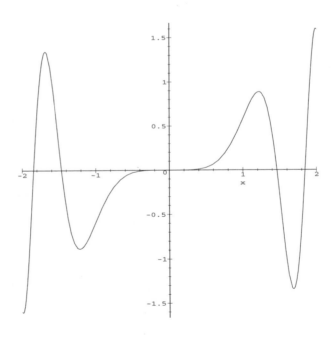

Chapter 2

Integers and rationals

This chapter is devoted to the arithmetic of integers and rationals. At the start, we use Maple just as a calculator. The basic elements of the language are introduced as they become necessary to illustrate arithmetical concepts.

2.1   Integers

The building blocks of arithmetic are the *natural numbers,* which form a *set* denoted by $\mathbb{N}$

$$\mathbb{N} = \{0, 1, 2, 3, \ldots\}.$$

The notion of a set — a collection of distinct objects — is fundamental in mathematics. It will be dealt with in some detail in later chapters. Here it suffices to say that the objects constituting the set $\mathbb{N}$ are the integer zero and the positive integers, which should be familiar to everyone. These elements are listed within curly brackets, separated by commas. As we cannot list them all, we resort to dots.

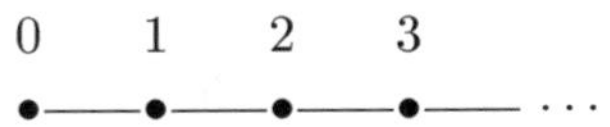

$$
\begin{array}{cccc}
0 & 1 & 2 & 3
\end{array}
$$

Figure 2.1: Graphical representation of the set $\mathbb{N}$ of natural numbers, as evenly spaced points on a semi-infinite line.

To say that an integer n belongs to $\mathbb{N}$, or is an *element* of $\mathbb{N}$, we use the symbol $\in$. Its negation is $\notin$. Thus,

$$23 \in \mathbb{N} \qquad\qquad -1 \notin \mathbb{N}.$$

The symbol $\mathbb{N}^+$ will be used to denote the set of *positive* integers

$$\mathbb{N}^+ = \{1, 2, 3, \ldots\}.$$

As an arithmetical environment, the set $\mathbb{N}$ is plainly inadequate. For instance, 2 and 3 belong to $\mathbb{N}$, and so do their sum $2 + 3 = 5$ and product $2 \cdot 3 = 6$. But their *difference* $2 - 3 = -1$ does not. One says that $\mathbb{N}$ is *not closed under subtraction*, that is, subtraction may take you out of the set. The lack of closure under subtraction — a serious problem — can be stated without ever mentioning the word 'subtraction' or introducing the minus sign. It suffices to note that the equation

$$x + a = b \qquad\qquad a, b \in \mathbb{N} \qquad\qquad (2.1)$$

cannot always be solved for x in $\mathbb{N}$. This formulation is useful, because it only makes use of minimal ingredients: the set $\mathbb{N}$ and an operation, called addition, that allows us to combine its elements together.

To solve this problem, we adjoin to $\mathbb{N}$ the negative integers, to obtain the set $\mathbb{Z}$ of all integers

$$\mathbb{Z} = \{\ldots, -2, -1, 0, 1, 2, \ldots\}.$$

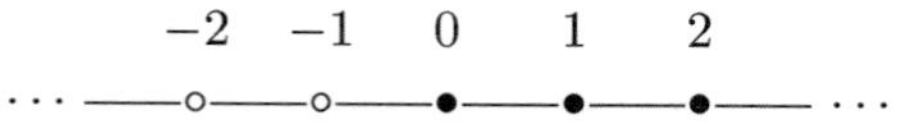

Figure 2.2: The set $\mathbb{Z}$, represented as points on a line. The white circles denote elements of $\mathbb{Z}$ which are not in $\mathbb{N}$.

Note that the elements of $\mathbb{Z}$ originate precisely from the distinct solutions x of equation (2.1). For instance, by setting $b = 0$ in (2.1) and letting a vary among all positive integers, we obtain all the negative integers. Likewise, by setting $a = 0$ and letting b vary in $\mathbb{N}$, we recover all elements of $\mathbb{N}$. It follows that every element of $\mathbb{N}$ is also an element of $\mathbb{Z}$, which we express by saying that $\mathbb{N}$ *is contained* in $\mathbb{Z}$ (or that $\mathbb{N}$ is a *subset* of $\mathbb{Z}$) and write

$$\mathbb{N} \subset \mathbb{Z}. \qquad\qquad (2.2)$$

This is called an *inclusion relation*. By the same token, $\mathbb{N}^+ \subset \mathbb{N}$.

The sum, difference, and product of any two elements of $\mathbb{Z}$ yield another element of $\mathbb{Z}$, and for this reason the arithmetic in $\mathbb{Z}$ is simpler than that in $\mathbb{N}$. In other words, each of the following equations

$$x = a + b \qquad x + a = b \qquad x = a \cdot b; \qquad a, b \in \mathbb{Z} \qquad (2.3)$$

can be solved by some $x \in \mathbb{Z}$. There is still a problem, though: the integers are not closed under *division* — the equation $x \cdot a = b$ cannot be added to the list (2.3). Indeed, -2 and 3 are integers, but $-2/3$ is not. This problem will lead to the introduction of the rational numbers.

Apart from the existence of arithmetical operations, two important structures are defined in $\mathbb{Z}$ which justify the representation of the integers as equally spaced points on a line, as in figure 2.2. They are an *ordering relation* and a *distance*.

First, the integers are *ordered*, in the sense that, for any $a, b \in \mathbb{Z}$, at least one of the following relations is satisfied:

$$a \leq b \qquad\qquad b \leq a. \qquad\qquad (2.4)$$

If both are satisfied, then $a = b$, while $a \not\leq b$ is equivalent to $a > b$, etc. The reader should be familiar with the following properties of the ordering relation $\leq$:

if	$a \leq b$ and $b \leq c$	then	$a \leq c$
if	$a \leq b$	then	$a + c \leq b + c$
if	$a \leq b$ and $0 \leq c$	then	$ac \leq bc$.

Second, we can measure the *distance* between two integers. To do so, we first define the *size* of an integer x as the absolute value of x, given by

$$|x| = \begin{cases} x & x \geq 0 \\ -x & x < 0. \end{cases} \qquad\qquad (2.5)$$

The quantity $|x|$ is non-negative, and is equal to zero precisely when $x = 0$. The numbers x and $-x$ have the same absolute value.

The distance between two integers x and y is defined as the size of the integer $x - y$, which is $|x - y|$. In particular, $|x| = |x - 0|$ is the distance between x and 0. From the definition, it follows that (*i*) the distance between x and y is the same as that between y and x (since $x - y$ and $y - x$ have the same size); (*ii*) the distance between x and y is zero precisely when $x = y$ (since $x - y$ is equal to zero only in this case); and (*iii*) the minimal positive distance between two integers is 1. Moreover, any three integers x, y, and z satisfy the inequality

$$|x - z| \leq |x - y| + |y - z| \qquad\qquad (2.6)$$

which is called the *triangle inequality*. The latter is justified by noting that the direct journey from x to z cannot be longer than that with an intermediate stop at y.

Figure 2.3: The triangle inequality (2.6). (*i*) The equal sign holds when y lies between x and z (including the endpoints). (*ii*) Strict inequality holds when z is in the middle.

2.2   Arithmetical expressions

Arithmetic in $\mathbb{Z}$ is straightforward with Maple.

> 5+3;

$$8$$

The *arithmetical expression* 5+3 consists of the *arithmetical operator* + and
the two *operands* 5 and 3. Its *value* is 8. The only syntactical constraint is
the semicolon at the end of the expression, which informs Maple that you
have finished typing in your input. Maple won't work until you put one,
and then press the RETURN key. If you do forget to put the semicolon,
that is, you type

> 5+3

followed by RETURN, then Maple will do nothing. After a while, you'll
begin to suspect that something is wrong, but do not panic, you can add
the semicolon at any time, followed by RETURN. One may perform several
additions at once

> 1+10+100+1000+10000+100000+1000000+10000000;

$$11111111$$

What happens if one makes a syntactical error? For instance, the expression
5++3 is illegal

> 5++3;

syntax error: '+' unexpected

When a syntactical error occurs, Maple writes an error message and po-
sitions the cursor immediately before the first incorrect character. In this
eventuality, just edit the incorrect character. Since Maple diagnoses one
error at a time, clearing all errors in an expression may take several steps.

The characters +, -, *, and / denote the four arithmetical operators:
addition, subtraction, multiplication, and division, respectively. An expression may contain any number of operators, but this creates a problem of
representation. For instance, the meaning of the Maple expressions 8+4+2
and 8*4*2 is clear. But what does 8/4/2 mean? If we translate it literally
into mathematical notation, we get a rather bizarre result

$$\frac{\frac{8}{4}}{2}$$

The ambiguity arises because division — unlike addition and multiplication
— is *non-associative*: that if a, b, and c are integers, with b and c nonzero,
then $(a/b)/c$ (meaning that a/b is done first, and then the result is divided
by c) is generally different from $a/(b/c)$ (meaning that b/c is done first, and
then a is divided by it). Thus, $(8/4)/2 = 2/2 = 1$ while $8/(4/2) = 8/2 = 4$.
Note that subtraction suffers from the same problem, while, as we have

pointed out, addition and multiplication are associative (see section 12.1 for formal definitions)

$$(a + b) + c = a + (b + c) \qquad (a \cdot b) \cdot c = a \cdot (b \cdot c).$$

It turns out that Maple always starts from the division on the *left*
> 8/4/2;

$$1$$

and to give priority to the rightmost division, parentheses are needed
> 8/(4/2);

$$4$$

The use of parentheses, even when redundant, often clarifies the meaning of an expression, so it is good practice to write (8/4)/2 instead of 8/4/2. Moreover, when it comes to redundant parentheses, Maple is very tolerant
> (((((8/((((4)))))))))/((2));

$$1$$

(Note that only *round* brackets are allowed: square and curly brackets have a specific meaning in Maple, and cannot be used to parenthesise expressions.) Sometimes Maple will insist that parentheses be used, even if the ambiguity could, in principle, be resolved. For instance, we cannot write $3 \cdot (-2)$ without parentheses, even though this expression has only one meaningful interpretation, which yields the value -6. If you type 3*-2 you get an error termination.

The exponentiation operator is represented by the character '^', that is, 2^14 means 2^{14}. The exponentiation is non-associative, but unlike for division or subtraction, Maple does require that a chain of exponentials be fully parenthesised, to resolve any ambiguity
> 2^(2^3);

$$256$$

> (2^2)^3;

$$64$$

If you type 2^2^3 you get an error termination.

That Maple is no ordinary calculator will become clear as soon as you try something like this

```
> 2^1279-1;
```

10407932194664399081925240327364085538615262247266704805319 1\
123504036080596733602980122394417323241848424216139542 81\
00779138356624832346490813990660567732076292412950938922\
034577318334966158355047295942054768981121169367714754 84\
7886696250138443826029173234888531116082853841658502825 5\
604666224831890918801847068222203140521026698435488732 95\
8028878050869736186900714720710555703168729087

$$(2.7)$$

The integer $2^{1279} - 1$ has 386 decimal digits. (It turns out to be a *prime*. To construct even larger primes, see the end of section 2.6). As the result does not fit on a single line, Maple has automatically appended the continuation character '\ ' which indicates that the output continues on the following line. In principle, we can compute and display extremely large integers, with up to half a million decimal digits, on most 32-bit computers. In practice, however, to compute and display very large integers, the computer may take longer than the maximum time we are prepared to wait (100 years, for instance). (To find out the maximum number of decimal digits supported by your computer, type `kernelopts(maxdigits)`.)

The expression 2^1279-1 contains two distinct arithmetical operators, exponentiation and subtraction. Again, there are two possibilities, namely

$$(a) \qquad 2^{1279} - 1 \qquad\qquad (b) \qquad 2^{1279-1} = 2^{1278}$$

depending on whether exponentiation (case (a)) or subtraction (case (b)) is done first. Because the result was an odd number, we tentatively conclude that Maple has given priority to exponentiation. But we do not know Maple's rule yet, since Maple could have given priority to the operator on the *left*. To settle the issue we try again

```
> 3-2^5;
```

$$-29$$

No, Maple really executes exponentiation before subtraction; otherwise, we would have obtained $(3 - 2)^5 = 1^5 = 1$.

```
> (3-2)^5;
```

$$1$$

If there is more than one operator, the order of evaluation of an expression is (i) exponentiation, followed by (ii) multiplication or division, followed by (iii) addition, subtraction, or negation (changing sign). If more than one operator with the same priority appears, evaluation proceeds from left to right for groups (ii) and (iii). Chains of exponentiation operators are illegal, as we have seen. The parentheses '(' and ')' may be used to alter the order of evaluation and to introduce signed operands. A sub-expression enclosed in parentheses is always evaluated first, and if there is more than one of them, with left to right priority.

Example 2.1. Let us construct the following expression

$$\frac{2^{3 \cdot 4 - 5} + 6}{7 \cdot 8} = \frac{(2^{(3 \cdot 4 - 5)} + 6)}{(7 \cdot 8)} = \frac{((2^{((3 \cdot 4) - 5)}) + 6)}{(7 \cdot 8)}. \tag{2.8}$$

This expression is normally written in the form appearing on the left, without parentheses. Its architecture — a numerator with some material at an exponent, and a denominator — implies three sets of parentheses (middle expression), while a further layer of parentheses spells, in detail, the order of evaluation of all arithmetical operators (rightmost expression).

In all, we have seven operands and six operators, and the order in which the operations have to be performed is the following (first a, then b, etc.)

$$2 \;\hat{}\; 3 * 4 - 5 + 6 \,/\, 7 * 8$$
$$c \quad a \quad b \quad d \quad f \quad e$$

Accordingly, we build the Maple expressions step by step, as follows:

```
2   ^ 3  * 4  -  5  + 6 / 7  * 8
2   ^ (3 * 4) -  5  + 6 / 7  * 8
2   ^((3 * 4) -  5) + 6 / 7  * 8
(2  ^((3 * 4) -  5)) + 6 / 7  * 8
((2 ^((3 * 4) -  5)) + 6) / 7  * 8
((2 ^((3 * 4) -  5)) + 6) / (7 * 8)
```

Thus, the correct expression is `((2^((3*4)-5))+6)/(7*8)` (the rightmost expression in (2.8)). By taking advantage of the order in which Maple performs arithmetical operations, we may simplify it to `(2^(3*4-5)+6)/(7*8)`, but no further.

Exercises

Exercise 2.1. Construct the following expressions:

1) $\dfrac{a}{b\,c}$ [1]   2) a^{b+c} [1]   3) $a^b - \dfrac{c}{d}$ [0]

4) $\dfrac{a^{bc} - d}{e}$ [2]   5) $\dfrac{a-b}{c-d}$ [2]   6) $\dfrac{a}{b + \dfrac{c}{d}}$ [1]

Try to minimize the use of parentheses by taking advantage of the order of evaluation of arithmetical operators. The minimal number of parentheses needed in each case is indicated in square brackets.

Exercise 2.2. Evaluate the following arithmetical expressions, trying to minimize the use of parentheses.

$$1 - \frac{1}{1766319049^2 - 61 \cdot 226153980^2} \qquad 3^3 \cdot 5 - \frac{3}{4 + \dfrac{2^{3^2-7} - 5}{17 \cdot 8}} \cdot 181.$$

The symbol '·' denotes multiplication. The expression on the left is an arithmetical miracle. Change just a digit in one of the integers in the denominator, and compare the result.

2.3   Some Maple

In this section we develop a repertoire of Maple tools.

Basic tricks

Blanks can be inserted freely between operators, operands, and parentheses, and because a new line character is interpreted as a blank, a command can be entered on several lines.

```
> 5
>            +        8
>
> ;
```

$$13$$

(This is why Maple does not complain when you forget the semicolon!) However, blanks are not allowed between the digits of a number, and in order to break long numbers across several lines of input, one must use the backslash character '\ '

```
> 1234\
> 56789 + 876543\
> 210;
```

$$999999999$$

Omitting the backslash causes an error termination. For instance, the statements

```
> 12
> 34+1;
```

are interpreted as 12 34+1, which is not a valid expression.

   The character '%', called a *ditto variable,* can be used in any expression to represent the value of the previous expression, that is, the output of the previous command.

```
> 51^2+80^2-1;
```

$$9000$$

```
> %/1000;
```

$$9$$

Here the ditto variable assumes the value of the expression `51^2+80^2-1`. Because it is the *value* that is substituted, and not the expression itself, the parentheses are not needed. In other words, the above expression is equivalent to `(51^2+80^2-1)/1000` and not to `51^2+80^2-1/1000` (verify this).

Maple supports other ditto variables. The variable '`%%`' assumes the value of the next-to-last command, and '`%%%`' assumes the value of the command before (but four or more percent signs mean nothing). Ditto variables are very handy when building up complicated expressions, since the correctness of each sub-expression can be checked separately. In the following example, we evaluate the expression $(12^3 + 1^3) - (10^3 + 9^3)$ in three steps

```
> 12^3+1^3;
```

$$1729$$

```
> 10^3+9^3;
```

$$1729$$

```
> %%-%;
```

$$0$$

When a computation involves a sequence of commands, one may wish to suppress the display of intermediate results. This is done by appending at the end of the command a colon ':' rather than a semicolon ';'

```
> 12^3+1^3:
> 10^3+9^3:
> %%-%;
```

$$0$$

Example 2.2. Evaluate the following arithmetical expression

$$((((1 + 1) \cdot 2 + 1) \cdot 3 + 1) \cdot 4 + 1) \cdot 5.$$

This expression is *nested,* and it should be built from the inside out

```
> 1:
> (%+1)*2:
> (%+1)*3:
> (%+1)*4:
> (%+1)*5;
```

$$325$$

In Maple, one can place more than one statement on the same line. All syntactical rules described above apply.

```
> 1+1:%-2;%%;
```

$$0$$

$$2$$

This feature should be used sparingly, because it can lead to illegible codes. It is also possible to display more than one result on the same line of output, by separating expressions with commas

```
> 1+1:%-2,%;
```

$$0, 2$$

The composite expression `%-2,%` constitutes a single statement (this construct will be considered in greater detail in a later chapter). Compare carefully the use of ditto variables in the last two examples.

Comments can be inserted in a Maple command using the character `#`. Everything after that character is ignored. Comments can also be inserted in the output, by enclosing them between *double* quotes.

```
> "one plus one equals",1+1; # this should be about two
```

$$one\ plus\ one\ equals,\ 2$$

```
> # cool!
```

Variables

Mathematical variables represent 'indeterminates' or 'unknowns'. So by $(a + b)^2$ we denote the square of the sum of two unspecified quantities. Maple operates much in the same way

```
> (a+b)^2;
```

$$(a + b)^2$$

We shall still refer to $(a + b)^2$ as the *value* of Maple's expression `(a+b)^2`. Expressions involving indeterminates are referred to as *algebraic* or *symbolic* expressions (as opposed to *arithmetical*).

When a variable is created, it has the status of an indeterminate (it is 'unassigned'). However, a variable can be assigned a value. The simplest values are numerical constants

```
> c:=3;
```

$$c := 3$$

The above statement is called an *assignment statement*. The *assignment operator* is ':=', and not the equal sign '=', which in Maple has a different meaning. Forgetting the colon before the equal sign in an assignment statement is a classic beginner's mistake, much like forgetting the semicolon at the end of an expression. Variables can also be assigned the value of arithmetical and algebraic expressions (and indeed of any valid Maple expression)

```
> c:=10^10;
```

$$c := 10000000000$$

```
> c:=(a+b)^2;
```

$$c := (a + b)^2$$

```
> %%,%;
```

$$10000000000, (a + b)^2$$

One sees that the *value* of an assignment statement, captured by the ditto variables, is equal to the value of the expression being assigned, and not to the assignment itself. This is because the assignment statement is *not* an expression, and so it does not have a value.

Once a variable is assigned a value, the latter is substituted whenever the variable's symbolic name appears in an expression

```
> a:=3:
> 2-a^a,c;
```

$$-25, (3 + b)^2$$

To return the variable to the status of indeterminate, we must use the *right* quotes as follows

```
> a:='a';
```

$$a := a$$

```
> 2-a^a,c;
```

$$2 - a^a, (a + b)^2$$

This can also be done temporarily within an expression

```
> z:=11^11;
```

$$z := 285311670611$$

```
> z+'z';
```

$$285311670611 + z$$

WARNING. A variable cannot be defined in terms of itself (*recursive definition*). Thus, if the variable `danger` was not previously assigned a value, the statement

```
> danger:=danger+1;
```

generates a warning message that a recursive definition has just been made. At the first attempt of evaluation, such as

```
> danger;
```

Maple will try to determine the value of the right-hand side of the assignment statement, which is the expression `danger+1`. To do so, Maple first

needs to evaluate **danger**, but because the name **danger** evaluates to itself, Maple will be referred repeatedly to the same assignment, until the
computer's memory is full and Maple grinds to a halt!

If, on the other hand, a variable has been previously assigned a value,
expressions of this type are perfectly legitimate

> z:=3:
> z:=z+1;

$$z := 4$$

In this case, the name **z** in the expression **z+1** evaluates to 3 rather than to
itself, so that the expression **z+1** evaluates to 4, which is the value assigned
back to **z**.

Maple allows great freedom in choosing variables' names. A name will
normally start with a letter, possibly followed by letters, digits and un-
derscores, totalling up to over half a million characters! Long, descriptive
names are commonplace in modern programming style

> first_square:=15140424455100^2;

$$first\_square := 229232452680590131916010000$$

> second_square:=158070671986249^2;

$$second\_square := 24986337342184324378845090001$$

> 109*first_square-second_square;

$$-1$$

Note the use of the underscore to join words. The same effect can be
achieved by means of upper-case letters, e.g., **FirstSquare**. It must be remembered that Maple distinguishes between upper- and lower-case letters.
So the names **Aa**, **aA**, **aa** and **AA** are all different.

Maple has *reserved words,* which have a pre-defined meaning, and which
should not be used for a different purpose. Some of them will not be
accepted as valid names (such as **and, stop, from**), while others are names
of Maple built-in functions which it would be unwise to overrule (such as
the names of the trigonometric functions **sin, cos, tan**, etc.).

Substitutions

Assigning a value to a variable by means of the assignment operator ':='
is not always desirable. If the assignment is temporary, it is often more
convenient to use the substitution function **subs**

> s:=x+y+x^2+y^2+x*y;

$$s := x + y + x^2 + y^2 + xy$$

> subs(x=100,y=99,s);

$$29900$$

```
> s;
```

$$s := x + y + x^2 + y^2 + xy$$

Substitution requires the operator '=' rather than the assignment operator. Note that x and y have been assigned a value with a single command, and that after the substitution, the variable s still retains its value, so it can be used again (even in the same expression)

```
> subs(x=1,s)-subs(y=z^2,s);
```

$$2 + 2y + y^2 - x - z^2 - x^2 - xz^2 - z^4 \tag{2.9}$$

Assigning values to x and y by means of the assignment operator would cause the original value of s to be temporarily lost. Its recovery would require clearing variables, and constructs of the type (2.9) could not be performed.

```
> x:=100:y:=99:s;
```

$$29900$$

```
> x:='x':y:='y':s;
```

$$x + y + x^2 + y^2 + xy$$

If more than one substitution is performed with a single **subs** command, Maple will proceed left to right in the substitution list. This form of multiple substitution is called *sequential*. In general, the order in which substitutions are performed matters, e.g.,

```
> subs(a=b,b=c,a+b+c);
```

$$3c$$

```
> subs(b=c,a=b,a+b+c);
```

$$b + 2c$$

(Study the above example with care.) In order to carry out a *simultaneous substitution*, it is necessary to group the substitution equations within curly brackets

```
> subs(a=b,b=c,c=a,a^3+b^2+c);
```

$$a^3 + a^2 + a$$

```
> subs({a=b,b=c,c=a},a^3+b^2+c);
```

$$b^3 + c^2 + a$$

The effect of the latter substitution is to permute the variables a, b, and c.

Example 2.3. Evaluate the following algebraic expression

$$4 \left(\frac{1}{a^2} - a \right) \left(\frac{1}{b^4} + b^2 \right) - \frac{9\,a}{\dfrac{1}{(b-1)^2} - b + 1}.$$

for $a = 2$ and $b = -1$. We build this expression starting from the sub-expression $1/a^2 - a$, using ditto variables and substitutions. Then we substitute numerical values for a and b.

```
> 1/a^2-a:
> 4*%*subs(a=-b^2,%)-9*a/subs(a=b-1,%):
> subs(a=2,b=-1,%):
```

$$-22$$

Relational expressions

We consider the problem of comparing the value of arithmetical expressions, with respect to the ordering relation (2.4) defined in $\mathbb{Z}$. For instance, suppose we want to decide which is the largest between 2^{1000} and 3^{600}. A straightforward way to do this is to take the difference between the two numbers and see whether it is positive, negative, or zero.

```
> 2^1000-3^600;
```

> 10715086071862654470207211642660131351594127758931911765961\
> 84728909219525443393780130210206303932798551557922601446741\
> 66052128348099701276136689688445564748966436835040418914761\
> 00477460737032825884980836183544889631284116847973632934411\
> 78560172497199203137042607882027712898454153704577866946131\
> 325762937375

Because this number is positive, we conclude that $2^{1000} > 3^{600}$. However, the above procedure is clearly unsatisfactory, since it requires generating all the digits of a very large number, when all we need is its sign.

Maple offers the possibility of obtaining a straight answer to questions such as 'is 2^{1000} greater than 3^{600}?'.

```
> evalb(2^1000 > 3^600);
```

$$true$$

The above expression features the *relational expression* 2^1000 > 3^600, which is evaluated with the Maple function `evalb`.

A *relational expression* is one that links numerical and logical data. It consists of two arithmetical expressions which are compared by means of an equality or an inequality, called a *relational operator*. If the relation is satisfied, the expression evaluates to the *Boolean* (or *logical*) constant *true* and to the constant *false* otherwise (unless Maple has difficulties in deciding, in which case it returns the value $FAIL$). The Maple relational operators are:

=	equal	>=	greater or equal	>	greater
<=	less or equal	<	less	<>	not equal

To evaluate a relational expression to a Boolean value, we use the function evalb (evaluate to Boolean). Explicit evaluation is needed here because such expressions may also occur as equations or inequalities, which can be manipulated algebraically

```
> evalb(-3<-2), evalb(-3=-2);
```

$$true, \, false$$

```
> 1+1=z;x<-3*y^2;
```

$$2 = z$$
$$x < -3\,y^2$$

```
> %%+%;
```

$$2 + x < z - 3\,y^2$$

We apply the above construct to the problem of comparing distances in $\mathbb{Z}$ with Maple. To measure a distance, we make use of the absolute value function (2.5), which is implemented in Maple by the function abs

```
> abs(-12),abs(53*9100^2-66249^2);
```

$$12, \, 1$$

The combined use of evalb and abs is illustrated in the following example.

Example 2.4. Let $n = 100^{100}$. Which of $n_1 = 101^{99}$ or $n_2 = 99^{101}$ is closer to n? To find out, we establish the truth or falsehood of the inequality

$$|n - n_1| < |n - n_2|\,.$$

```
> n:=100^100:n1:=101^99:n2:=99^101:
> evalb(abs(n-n1) < abs(n-n2));
```

$$true$$

So n_1 is the closer of the two to n.

Exercises

Exercise 2.3. By means of some experiments, verify the rules concerning the order of evaluation of arithmetical operators.

Exercise 2.4. In this exercise you should use the function subs.
(a) Construct the following algebraic expression

$$1 + z + z^2 + z^3 + z^4,$$

whence evaluate the sum

$$1 + 2 + 2^2 + 2^3 + 2^4 - 1 - 3 - 3^2 - 3^3 - 3^4 + 1 - 4^2 + 4^4 - 4^6 + 4^8.$$

(*b*) Construct the following algebraic expression

$$\frac{1 + x + x^2 + x^3}{1 - y - y^2 - y^3},$$

whence evaluate the arithmetical expression

$$\frac{1 + 3 + 3^2 + 3^3}{1 - 2 - 2^2 - 2^3} \cdot \frac{1 + 2 - 2^2 + 2^3}{1 - 3 + 3^2 - 3^3}.$$

Exercise 2.5. In this exercise you should not display any digit. Display only the minimal output required to establish your result.

(*a*) Place the following integers in ascending order

$$100^{100}, \ 80^{120}, \ 60^{140}, \ 40^{160}, \ 20^{180}.$$

(*b*) Find the largest integer whose 17th power is smaller than 10^{30}.

2.4 Divisibility

Divisibility is one of the most important concepts in arithmetic. We begin with some definitions.

Given two integers d and n, we say that d *divides* n (or d is a divisor of n, or n is a multiple of d) if there exists an integer q such that $n = dq$. Such q is called the *quotient* of the division of n by d. With reference to (2.3) and following comments, we see that divisibility characterizes the extent to which the missing equation $x \cdot a = b$ can be solved.

If d divides n we write $d \mid n$, and if d does not divide n we write $d \nmid n$. Thus, $3 \mid 21$ because $21 = 3 \cdot 7$, while $12 \nmid 21$ because the equation $21 = 12 \cdot q$ has no integer solution q. An integer is *even* if it is divisible by 2, and *odd* if it is not. Two integers have the *same parity* if they are both even or both odd. If one is even and the other is odd, they have *opposite parity*.

Since for all n we have $n = 1 \cdot n$, it follows that 1 and n divide n, and since $0 = d \cdot 0$, we see that any integer divides 0. Also if d divides n, so does $-d$. For this reason, in issues involving divisibility it is customary to consider the positive divisors only. A *proper divisor* d of n is a divisor which is neither equal to 1 nor to n.

Example 2.5. The number 0 has infinitely many divisors, 12 has six divisors (1,2,3,4,6, and 12), four of which are proper, 11 has two divisors (1 and 11), and 1 has one divisor.

Divisibility has a simple geometrical interpretation. If d divides n, we can arrange n points on the plane to form a rectangular array with d rows. Interchanging rows and columns, that is, rotating the array by a right angle, yields another array with $q = n/d$ rows. This geometrical argument shows that *divisors come in pairs*, that is, if d divides n then also n/d

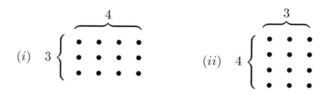

Figure 2.4: Graphical representation of the divisibility of $n = 12$ by (i) $d = 3$, and (ii) by its twin divisor $n/d = 12/3 = 4$.

divides n and vice versa. The above symmetry has an arithmetical counterpart: if d divides n, the role of d and $q = n/d$ in the equation $n = dq$ is interchangeable.

Because every divisor d of n has the twin divisor n/d, can we conclude that the number of divisors of an integer is always even? Not necessarily, because d and n/d may coincide, giving $n = d^2$, a *square*. Thus, an integer is a square precisely when it has an *odd* number of divisors.

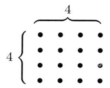

Figure 2.5: The integer $n = 16 = 4^2$ is a square, and the divisor $d = 4$ coincides with its twin $n/d = 16/4 = 4$.

This pairing of divisors has an important consequence. Suppose that d is a divisor of n such that $d^2 \leq n$ (that is, $d \leq \sqrt{n}$). Then we have $n/d^2 \geq 1$, whence $(n/d)^2 = n(n/d^2) \geq n$. This shows that the twin divisor n/d of d satisfies the inequality $(n/d)^2 \geq n$ (or $n/d \geq \sqrt{n}$). So in order to find all divisors of n, we must test for divisibility only those integers d for which $1 \leq d^2 \leq n$ (that is, $1 \leq d \leq \sqrt{n}$).

Example 2.6. To find the divisors of 30, we must test for divisibility all integers up to and including 5. The divisors of 30, regrouped in pairs, are $(1, 30)$, $(2, 15)$, and $(3, 10)$. The divisors of $36 = 6^2$ are $(1, 36)$, $(2, 18)$, $(3, 12)$, $(4, 9)$, and 6. Because 36 is a square, one of its divisors has no twin, and the number of divisors of 36 is odd.

Quotient and remainder

Given two natural numbers d and n, with $d \neq 0$, one can always find two uniquely determined natural numbers q and r such that

$$n = dq + r \qquad 0 \leq r < d. \qquad (2.10)$$

The integers q and r are called the *quotient* and the *remainder* of the division of n by d, respectively. The existence of such integers with the stated properties can be established as follows. We arrange n points on the plane to form a pattern which is as close as possible to a rectangular array with d rows. In other words, we maximize the number of columns with d elements, as illustrated below for the case $n = 14$ and $d = 3$.

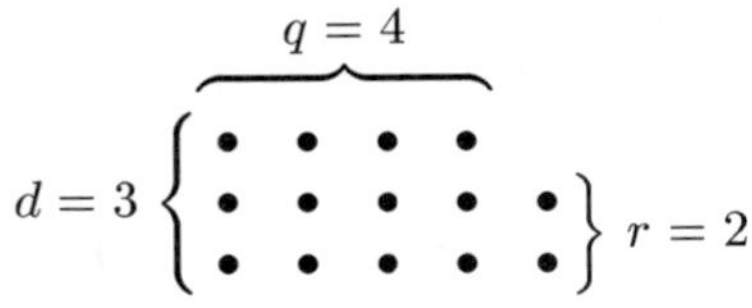

Figure 2.6: Dividing 14 by 3 gives quotient 4 and remainder 2.

Let q be the number of complete columns (d elements). Then there exists, at most, one incomplete column, whose number r of elements is necessarily smaller than d (otherwise, we could form more complete columns). This establishes the equation and the inequality in (2.10). The condition that r be smaller than d is what makes the definition of remainder interesting. In particular, it follows that d divides n precisely when the corresponding remainder is zero, in which case the $n = dq$ points form a rectangular array with d rows.

Note that one speaks of the quotient of the division of n by d whether or not d divides n. When d divides n, the definition (2.10) with $r = 0$ reduces to that given at the beginning of this section.

The greatest common divisor

A *common divisor* of two integers is an integer that divides both. Thus, if $n = dq$ and $n' = dq'$, then d is a common divisor of n and n'.

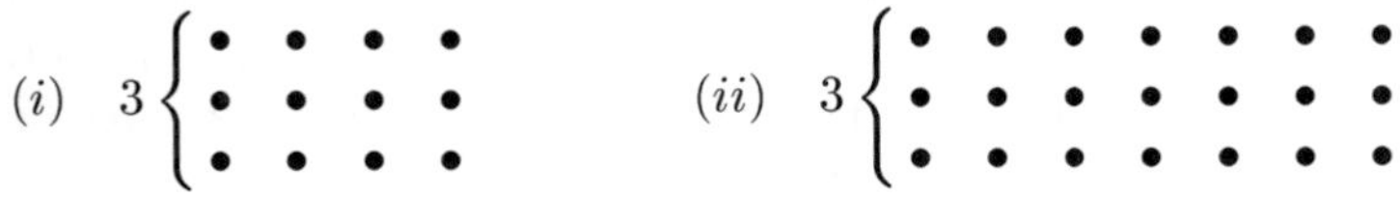

Figure 2.7: The integers (*i*) $12 = 3 \cdot 4$ and (*ii*) $21 = 3 \cdot 7$ have the common divisor 3.

If d divides n and n', then d also divides $n \pm n'$. Indeed, from $n = dq$ and $n' = d'q$, we have $n \pm n' = d(q \pm q')$. (This is also clear from the above figure.)

The *greatest common divisor* of two integers x and y, denoted by $\gcd(x, y)$, is the largest among the common divisors of x and y.

$$\gcd(-15, 21) = 3 \qquad \gcd(1, -7) = 1 \qquad \gcd(0, 7) = 7.$$

By definition, $\gcd(x,y) = \gcd(|x|,|y|)$. Also, $\gcd(x,0) = |x|$, and, for convenience, $\gcd(0,0)$ is defined to be zero. Thus, $\gcd(x,y) \geq 0$. The definition of greatest common divisor may be extended to the case of more than two integers in an obvious fashion. Thus, $\gcd(x_1,\ldots,x_n)$ is the largest among the common divisors of the n integers $x_1,\ldots,x_n$.

The integers $x_1,\ldots,x_n$ are said to be *relatively prime*, or *coprime*, if their greatest common divisor is equal to 1. For example, $14,15$ are relatively prime, and so are $10,15,18$. The latter example shows an interesting phenomenon. We have $\gcd(10,15,18) = 1$, in spite of the fact that $\gcd(10,15) > 1$, $\gcd(10,18) > 1$ and $\gcd(15,18) > 1$. So when n is greater than 2, n integers may be relatively prime without being so when considered in pairs. We say that n integers are *pairwise relatively prime* if any two of them are. For instance, $26,33,35$ are pairwise relatively prime. Pairwise relative primality is stronger than relative primality, in the sense that the former implies the latter (see exercises).

The *least common multiple* of two nonzero integers x and y, denoted by $\mathrm{lcm}(x,y)$, is the smallest positive integer divisible by x and y. For any integer x we define $\mathrm{lcm}(x,0) = 0$. For instance, $\mathrm{lcm}(12,21) = 84$. As for the greatest common divisor, the definition of least common multiple can be extended to the case of more than two integers.

Maple functions for divisibility

Maple implements a large number of arithmetical functions, including all elementary functions related to divisibility.

The functions `iquo(n,d)` and `irem(n,d)` return the quotient and the remainder, respectively, upon dividing n by d. (The prefix i means 'integer'. Maple also has the functions `quo`, `rem` which carry out the corresponding operations for polynomials, as we shall see.) These functions accept also negative arguments, in which case the definition of quotient and remainder given in (2.10) is generalized as follows

$$n = dq + r \qquad 0 \leq |r| < |d|, \qquad n\,r \geq 0. \qquad (2.11)$$

The behaviour of `iquo` and `irem` when one or both arguments are negative is best illustrated with examples.

```
> iquo(23,7),iquo(23,-7),iquo(-23,7),iquo(-23,-7);
```

$$3, -3, -3, 3$$

```
> irem(23,7),irem(23,-7),irem(-23,7),irem(-23,-7);
```

$$2, 2, -2, -2$$

We see that the remainder is negative whenever n is negative, while the quotient is negative if n and d have opposite signs.

The function `igcd(x1,...,xn)` and `ilcm(x1,...,xn)` return the greatest common divisor and the least common multiple, respectively, of the integers `x1` through `xn`. For a summary of the main Maple arithmetical functions, see the last section of this chapter.

Example 2.7. *Divisibility and greatest common divisor.* There is a close relationship between divisibility of b by a and the greatest common divisor of a and b. Indeed, if $a > 0$, then a divides b precisely when $\gcd(a, b) = a$. To see this, observe that if $\gcd(a, b) = a$, then $a|b$, by definition of greatest common divisor. Conversely, let $a|b$. Because $a|a$, then a divides both a and b, and since a is the largest integer dividing a, it follows that $\gcd(a, b) = a$.

Example 2.8. *Divisibility and remainder.* We can use `irem` to test divisibility, since d divides n exactly when `irem(n,d)` is equal to zero, in which case `evalb(irem(n,d)=0)` evaluates to *true*. For instance, $2^{191} - 1$ is divisible by 383

```
> evalb(irem(2^191-1,383)=0);
```

$$true \hspace{8cm} (2.12)$$

Pythagorean triples*

A *Pythagorean triple* is a set of three positive integers (x, y, z) such that

$$x^2 + y^2 = z^2. \hspace{6cm} (2.13)$$

For instance, $(3, 4, 5)$ is such a triple. The elements of a triple are the lengths of the sides of a right triangle, from Pythagoras' theorem, and so the problem of finding Pythagorean triples amounts to finding right triangles whose sides all have integer length.

If (x, y, z) is a triple, then by multiplying equation (2.13) by d^2 we find that (dx, dy, dz) is another triple. So from any triple (x, y, z) one can generate infinitely many of them, by letting $d = 1, 2, 3, \ldots$. If $d > 1$ then the elements of such a triple are not relatively prime, having the common divisor d.

Conversely, if d is the greatest common divisor of x, y, and z, then d^2 can be cancelled out from equation (2.13) and one finds that $(x/d, y/d, z/d)$ is also a triple whose elements are relatively prime. A Pythagorean triple with this property is said to be *primitive.*

It turns out that all triples can be derived from primitive ones, so the primitive case is the most interesting one. The following algorithm for finding *all* primitive Pythagorean triples is due to Diophantus of Alexandria (about 250 a.d.).

Consider two positive integers p and q such that (i) $p > q$, (ii) p and q are relatively prime, (iii) p and q are of opposite parity (i.e., one is even and the other is odd). Then $(2\,p\,q,\, p^2 - q^2,\, p^2 + q^2)$ is a primitive Pythagorean triple. That this is actually a triple follows from the algebraic identity

$$(2\,p\,q)^2 + (p^2 - q^2)^2 = (p^2 + q^2)^2 \hspace{4cm} (2.14)$$

which is valid for all p and q. It is possible to prove that such a triple is primitive, and that all primitive triples can be obtained in this way (see exercises that follow).

The primitive triples for $p \leq 7$ are tabulated below

PYTHAGOREAN TRIPLES

p	q	x	y	z
2	1	4	3	5
3	2	12	5	13
4	1	8	15	17
4	3	24	7	25
5	2	20	21	29
5	4	40	9	41
6	1	12	35	37
6	5	60	11	61
7	2	28	45	53
7	4	56	33	65
7	6	84	13	85

Exercises

Exercise 2.6. Show that every common divisor of two integers divides their greatest common divisor.

Exercise 2.7. Show that if n integers are pairwise relatively prime, they are relatively prime.

Exercise 2.8. Find all primitive Pythagorean triples corresponding to $p = 987$ and $q \geq 970$ (five in all).

Exercise 2.9. Find a primitive Pythagorean triple whose elements all have at least 20 decimal digits (i.e., $x, y, z \geq 10^{19}$). Verify directly that your solution is indeed a primitive triple, that is, that x, y, and z satisfy equation (2.13) and are relatively prime.

Exercise 2.10*. Prove that if $p > q > 0$, $\gcd(p, q) = 1$, and p and q have opposite parity, then $x = 2pq, y = p^2 - q^2$, and $z = p^2 + q^2$ are *pairwise* relatively prime. This, along with (2.14), establishes that (x, y, z) is a primitive Pythagorean triple. (In this exercise, you will need the fundamental theorem of arithmetic — see below.)

Exercise 2.11**. Prove that any primitive triple can be obtained with Diophantus' algorithm.

2.5 Rationals

The operations of sum, subtraction, and multiplication can be performed unrestrictedly in $\mathbb{Z}$. And division? We have seen that if a and b are integers, then a/b is not necessarily an integer, and indeed the notion of divisibility is introduced precisely to handle this possibility. We say that $\mathbb{Z}$ is *not closed under division.*

To characterize the lack of closure solely in terms of the elements of $\mathbb{Z}$ and the operation of multiplication, we note that if a and b are integers, with $a \neq 0$, then the equation

$$x \cdot a = b \qquad a, b \in \mathbb{Z}, \quad a \neq 0 \qquad (2.15)$$

cannot always be solved for $x \in \mathbb{Z}$. Note the structural similarity between equations (2.1) and (2.15), with $\mathbb{N}$ and $+$ in the former playing the same role as $\mathbb{Z}$ and $\times$ in the latter.

To solve this problem, we adjoin to $\mathbb{Z}$ the solutions of the equations (2.15), which are called *rational numbers.* They form a set denoted by $\mathbb{Q}$. A rational number is the *ratio* of two integers (whence its name!)

$$x = \frac{b}{a} \qquad a, b \in \mathbb{Z}, \quad a \neq 0. \qquad (2.16)$$

If a divides b, then $b = aq$, and the number x in (2.16) is equal to q, an integer. By letting $a = 1$ in (2.15) and letting b run over $\mathbb{Z}$, we obtain all elements of $\mathbb{Z}$. This means that all integers are rational, and the inclusion relation (2.2) extends to

$$\mathbb{N} \subset \mathbb{Z} \subset \mathbb{Q}.$$

Clearly not every rational is an integer.

The arithmetical operations with rationals are defined from those in $\mathbb{Z}$ as follows

$$\frac{a}{b} \pm \frac{c}{d} = \frac{a \cdot d \pm b \cdot c}{b \cdot d} \qquad \frac{a}{b} \cdot \frac{c}{d} = \frac{a \cdot c}{b \cdot d} \qquad \frac{a}{b} \div \frac{c}{d} = \frac{a}{b} \cdot \frac{d}{c}$$

where we have assumed that $b \cdot d \neq 0$, and also that $c \neq 0$ in the last equation. Now sum, subtraction, multiplication, and division of elements of $\mathbb{Q}$ yield an element of $\mathbb{Q}$, with the only exception of division by zero. We have at last constructed a set where all four arithmetical operations can be performed unrestrictedly. A set with these properties is very special: it is called a *field* (see section 12.3).

The set of *positive* rational numbers is denoted by $\mathbb{Q}^+$. Thus, $\mathbb{Q}^+$ is the set of numbers x of the form (compare with (2.16))

$$x = \frac{b}{a} \qquad a, b \in \mathbb{N}^+.$$

The very definition of a rational number creates a problem of representation. Let $a = -6$ and $b = 4$. Infinitely many pairs of integers exist, which represent the same rational b/a

$$x = \frac{-2}{3} = \frac{2}{-3} = \frac{-4}{6} = \frac{4}{-6} = \frac{-6}{9} = \frac{6}{-9} = \cdots.$$

To obtain a unique representation we shall require that a be positive, and that a and b be relatively prime. A rational b/a satisfying these requirements is said to be in *reduced form*. Thus, the reduced form of $4/(-6)$ is $-2/3$. The reduced form affords the most economical representation of a rational number.

Maple always represents and displays rationals in reduced form

```
> 4/(-6);
```

$$\frac{-2}{3}$$

Maple provides the function `numer` and `denom` which return the numerator and the denominator of the value of a rational expression, respectively

```
> 22/8:%,numer(%),denom(%);
```

$$\frac{11}{4}, 11, 4$$

Rationals constitute a new type of object in Maple — a new *data type*. To find out what Maple calls them, we use the function `whattype`

```
> whattype(2),whattype(2/3);
```

$$integer,\ fraction$$

Example 2.9. The fact that Maple always represents rationals in reduced form yields a straightforward method to test divisibility, which consists of evaluating the expression a/b, and then checking whether or not the result is an integer

```
> 111111111111/1111:%,whattype(%);
```

$$100010001,\ integer$$

```
> 111111111111/11111:%,whattype(%);
```

$$\frac{11111111111}{11111},\ fraction$$

Thus, 1111 divides 111111111111 whereas 11111 does not.

Ordering and distance in $\mathbb{Q}$

The set $\mathbb{Q}$ inherits from $\mathbb{Z}$ ordering and distance, as defined in section 1.1. This means that we can decide about the ordering of any two rationals from an order relation involving only integers

$$\frac{a}{b} \leq \frac{c}{d} \quad \Longleftrightarrow \quad ad \leq bc \qquad bd > 0,$$

where the symbol '$\Longleftrightarrow$' means that the two relations on the opposite sides are equivalent — they are either both true or both false (see Chapter 3).

Similarly, we define the absolute value $|x|$ of a rational x, in terms of a ratio of absolute values of integers

$$\left|\frac{a}{b}\right| = \frac{|a|}{|b|}.$$

As we did for the integers, we then let the distance between x and y be $|x - y|$.

Let $x = b/a$ be a rational number. Then from (2.10) integers q and r can be found such that

$$b = qa + r \qquad 0 \le r < a.$$

Dividing out this equation by a, we obtain

$$x = \frac{b}{a} = q + \frac{r}{a}. \tag{2.17}$$

The integer q — the quotient of the division of b by a — is the *integer part* of x. The rational number r/a is called the *fractional part* of x, denoted by $\{x\}$. Thus

$$\frac{23}{7} = 3 + \frac{2}{7} \implies \left\{\frac{23}{7}\right\} = \frac{2}{7} \qquad \frac{-23}{7} = -4 + \frac{5}{7} \implies \left\{\frac{-23}{7}\right\} = \frac{5}{7}.$$

Note the asymmetry between positive and negative values of x. In particular, with reference to equation (2.17), we see that q is the largest integer not greater than x if x is positive, and the smallest integer not smaller than x if x is negative.

Example 2.10. Let x be a non-negative rational number. From (2.17) we see that the fractional part of x is given by `irem(numer(x),denom(x))/denom(x)`. (The Maple function `frac(x)` does the same.) For negative values of x both expressions give the fractional part in accordance with the more symmetrical equation (2.11) in place of (2.10)

```
> -23/7:frac(%);
```

$$-\frac{2}{7}$$

Example 2.11. Let $r_1 = 21/34$, $r_2 = 55/89$, $r_3 = 34/55$. Show that r_2 lies between r_1 and r_3. We use the triangle inequality (2.6)

```
> r1:=21/34:r2:=55/89:r3:=34/55:
> evalb(abs(r1-r3)=abs(r1-r2)+abs(r2-r3));
```

$$true$$

The minimal distance between two distinct integers is 1, and in particular, 1 is the smallest positive integer. We now show that, by contrast, there is no minimal distance between rationals, whence there is no smallest positive rational. This important property is a consequence of the *Archimedean property* of the rationals, stating that for all positive rationals x and y there exists a positive integer n such that $nx > y$. In words, a sufficiently large number of copies of any positive quantity can be made arbitrarily large. Now let $y = 1$ and let x be a nonzero rational. Then $|x| > 0$, and the

Archimedean property ensures that we can find a positive integer n such that $n\,|x| > 1$ or

$$\frac{1}{n} < |x|.$$

No matter how small x is, we have found a positive rational $1/n$ which is closer to zero than x. So there is no smallest positive rational.

In the above construction, the closer x is to zero, the larger n will be. This observation can be generalized: if two rationals are close to each other, then the product of their denominators must be large (and therefore at least one of them must have a large denominator). To prove this, let us assume that the distance between $x = a/b$ and $y = c/d$ is small (in particular, less than 1). Then there exists a positive integer n such that

$$|x - y| = \left|\frac{ad - cb}{bd}\right| = \frac{|ad - cb|}{|bd|} \leq \frac{1}{n}. \tag{2.18}$$

Because $x \neq y$, it follows that $ad - cb \neq 0$, and since all quantities are integers we have that $|ad - cb| \geq 1$. From this and the inequality in (2.18) we obtain

$$|bd| \geq |ad - cb|\, n \geq n. \tag{2.19}$$

This inequality shows that the product bd of the denominators of x and y is at least as large as n, in absolute value. From (2.18) we see that as x approaches y, we can let n approach infinity, and with it the product of the denominators, from (2.19).

A straightforward method for finding a rational close to a given rational is based on the following construction. For $n > 0$ we consider the set $\frac{1}{n}\mathbb{Z}$ of all rationals with denominator n. These are the numbers of the form k/n, for $k \in \mathbb{Z}$. The set $\frac{1}{n}\mathbb{Z}$ consists of equally spaced rationals with spacing $1/n$. Note that

$$\mathbb{Z} \subset \frac{1}{n}\mathbb{Z} \subset \mathbb{Q} \qquad n \geq 1.$$

For any choice of k, the element $k/n \in \frac{1}{n}\mathbb{Z}$ is separated from its two neighbouring elements $(k \pm 1)/n$ by the minimal distance $1/n$.

Let now $x = a/b$ be the given rational. Then, for every $m > 0$ we have $x = am/bm$, whence

$$\frac{a}{b} \in \frac{1}{bm}\mathbb{Z} \qquad m = 1, 2, \ldots.$$

The neighbours of x in $\frac{1}{bm}\mathbb{Z}$, given by

$$\frac{am \pm 1}{bm}$$

lie at distance $1/bm$ from x. So, by choosing m to be sufficiently large, we can make this distance be as small as we please.

While finding a rational close to a given rational is straightforward, the problem becomes much more subtle if we try at the same time to minimize the size of the denominator. For given n, the minimal denominators will occur when $ad - bc = \pm 1$ and when b and d are of similar magnitude, each approximately equal to the square root of n.

Example 2.12.   Find a rational lying at distance smaller than 10^{-4} from $x = 96/145$. We let $a = 96$, $b = 145$, and $n = 10^4$, and compute the smallest m for which $mb > n$.

```
> a:=96:b:=145:n:=10^4:
> m:=iquo(n,b)+1;
```

$$m := 69$$

A desired rational is then the right neighbour of x in $\frac{1}{bm}\mathbb{Z}$, given by

```
> (a*m+1)/(b*m);
```

$$\frac{1325}{10005}$$

But we can do better with a much smaller denominator, e.g.,

```
> c:=47:d:=71:
> abs(a/b-c/d);
```

$$\frac{1}{10295}$$

With reference to (2.19) note that

```
> abs(a*d-b*c);
```

$$1$$

(The rational 47/71 was found with a number-theoretical technique called *continued fractions.* See (2.22) for an example.)

Example 2.13.   Let a/b and c/d be two rational numbers, in reduced form. Their *mediant* is defined as

$$\frac{a+c}{b+d}. \tag{2.20}$$

We show that the mediant of two reduced rationals lies between them. Let us assume that $a/b < c/d$. Then, since bd and $d(b+d)$ are positive, we have

$$\frac{a}{b} < \frac{c}{d} \quad \Longleftrightarrow \quad ad < bc \quad \Longleftrightarrow \quad ad + cd < bc + cd$$

$$\Longleftrightarrow \quad (a+c)d < (b+d)c \quad \Longleftrightarrow \quad \frac{a+c}{b+d} < \frac{c}{d}.$$

The other inequality is proven similarly. Another prominent rational lying between a/b and c/d is their *midpoint*

$$\frac{1}{2}\left(\frac{a}{b} + \frac{c}{d}\right) = \frac{ad + bc}{2bd}. \tag{2.21}$$

The denominator of the midpoint is, in general, much larger than that of the mediant. For this reason, the mediant may be used as a simple device for identifying low-denominator rationals between two given rationals.

Exercises

Exercise 2.12. Evaluate the following rational expression, first by hand, then with Maple

$$\frac{1 + \frac{1}{3} \div \left(1 + \frac{1}{18} - \frac{2}{12}\right) - \frac{1}{4}}{\frac{4}{9} \times \left(\frac{5}{14} - \frac{2}{3} \times \frac{3}{56}\right) + \frac{10}{28}}$$

Exercise 2.13. Consider the following expression, called a *continued fraction*

$$a + \cfrac{1}{b + \cfrac{1}{a + \cfrac{1}{b + \cfrac{1}{a}}}} \tag{2.22}$$

(*a*) Construct it by means of a single Maple expression.
(*b*) By repeated use of the ditto variable %, construct it recursively, in four or five steps. [*Hint:* begin from the end.]
(*c*) Use the function **subs** to evaluate (2.22) for $a = b = 1$, and then for $a = 10$ and $b = 20$.

Exercise 2.14. Let $a = 5001999958$ and let $b = 5004400966$.
(*a*) Compute the greatest common divisor d of a and b.
(*b*) Verify the following equation and inequalities

$$a = bq + r \qquad 0 \le r < b$$

where q is the quotient and r is the remainder upon dividing a by b.
(*c*) Verify that Maple represents a/b in reduced form, that is, check that a/b is displayed as a'/b' where

$$a' = \frac{a}{d} \qquad\qquad b' = \frac{b}{d}$$

and d is as above. The integers a' and b' are extracted from the value of a/b by means of the Maple functions **numer** and **denom**, respectively.
(*d*) Find a proper divisor of a that does not divide b.

Exercise 2.15. Let $a = 10^{20} - 2^{10}$ and let $b = 2^{60} - 1$. Determine the integer which is closest to a/b.

Exercise 2.16. Let

$$x = \frac{89}{144} \qquad y = \frac{377}{610} \qquad z = \frac{987}{1597}.$$

(*a*) Consider the triangle inequalities

(*i*) $|x - z| \le |x - y| + |y - z|$ (*ii*) $|x - y| \le |x - z| + |z - y|$.

Verify that strict inequality holds in case (*i*), and equality holds in case (*ii*).

(*b*) Find a rational number r lying between x and z, which is closer to z than y is. The denominator of r should be as small as possible (there is a solution with a 3-digit denominator; can you find it?).

2.6   Primes

A positive integer n greater than 1 is said to be *prime* if it has precisely two divisors: 1 and n (or, equivalently, if it has no *proper* divisor). If this is not the case, we say that n is *composite*. Note that 1 is not considered prime, for convenience (see below). There are 25 primes less than 100. They are:

$$2, 3, 5, 7, 11, 13, 17, 19, 23, 29, 31, 37, 41, 43, 47,$$
$$53, 59, 61, 67, 71, 73, 79, 83, 89, 97. \tag{2.23}$$

Euclid, around 300 b.c., proved that there are infinitely many primes. The number of primes not exceeding x is denoted by $\pi(x)$. Thus, $\pi(1) = 0$, $\pi(11) = 5$, $\pi(100) = 25$. Here is a brief table of this very important function

x	$\pi(x)$
10	4
10^2	25
10^3	168
10^4	1,229
10^5	9,592
10^6	78,489
10^7	664,579
10^8	5,761,455
10^9	50,847,534
10^{10}	455,052,511

$$\tag{2.24}$$

The density of primes decreases slowly. While a quarter of the integers less than 100 are prime, the proportion of primes among the first 10 billion integers has dropped to less than 5%.

The arithmetic in $\mathbb{Z}$ is based on the following result.

Theorem 1 (the fundamental theorem of arithmetic). *Every integer greater than one can be expressed as a product of primes, and this factorization is unique up to the order of the factors.*

Thus, $12 = 2^2 \cdot 3$. You can see why 1 is not considered prime. If it were, then $12 = 1 \cdot 2^2 \cdot 3$ would be a *different* decomposition into primes. In general, the prime decomposition of an integer n, positive or negative, takes the form $(n \neq 0, \pm 1)$

$$n = \pm p_1^{e_1} \cdot p_2^{e_2} \cdots p_k^{e_k} = \pm \prod_{i=1}^{k} p_i^{e_i}, \qquad (2.25)$$

where we stipulate that the primes p_i are all distinct and that the exponents e_i are positive integers. The primes $p_1, \ldots, p_k$ are called the *prime divisors* of n.

If n has the prime decomposition (2.25), and $n = dq$, then the prime decomposition of dq is the same as that of n, from the fundamental theorem. So if d divides n, the prime divisors of d must also be prime divisors of n, and their exponent cannot exceed the corresponding exponent for n.

When constructing a divisor of n from (2.25), we find that there are $e_i + 1$ possible choices for each exponent e_i (from 0 to e_i). So the number $\sigma(n)$ of divisors of n is given by

$$\sigma(n) = (e_1 + 1)(e_2 + 1) \cdots (e_k + 1). \qquad (2.26)$$

Example 2.14. We have

$$84 = 2^2 \cdot 3^1 \cdot 7^1.$$

Thus, 84 has three prime divisors. In the notation of equation (2.25) we have $k = 3$, $p_1 = 2$, $p_2 = 3$, $p_3 = 7$, $e_1 = 2$, $e_2 = 1$, $e_3 = 1$. The number of divisors of 84 is $\sigma(84) = (e_1 + 1)(e_2 + 1)(e_3 + 1) = 3 \cdot 2 \cdot 2 = 12$.

Example 2.15. We determine the structure of the prime decomposition of all integers with 10 divisors. From (2.26) we have

$$(e_1 + 1)(e_2 + 1) \cdots (e_k + 1) = 10 = 2 \cdot 5.$$

We have two possibilities, namely $k = 1, e_1 = 9$, or $k = 2, e_1 = 1, e_2 = 4$. Thus, if n has 10 divisors, its prime decomposition must have the form

$$n = p_1^9 \qquad \text{or} \qquad n = p_1 \, p_2^4$$

where p_1 and p_2 are arbitrary distinct primes. From the infinitude of primes, it follows that the number of integers with 10 divisors is infinite.

Maple has a function called `ifactor` which returns the unique prime decomposition of an integer, as referred to in the fundamental theorem of arithmetic

```
> ifactor(3^52-2^52);
```

$$(5)\,(13)^2\,(53)\,(79)\,(1093)\,(13761229)\,(29927)\,(4057)$$

Note that Maple is rather untidy — the prime factors are not necessarily displayed in ascending order.

The existence of the fundamental theorem gives us a systematic way of constructing all divisors of any given integer n, once we have its prime decomposition. Obtaining the prime decomposition of an integer is conceptually easy. All that is needed is testing the divisibility of n by all primes $p \leq \sqrt{n}$. However, if n is large, this procedure is hopelessly slow. But even with the most sophisticated factorization algorithms (such as the one implemented by the function `ifactor`), factoring a large integer into primes remains an extremely difficult computational problem.

Example 2.16. The function `ifactor` affords a test for primality, albeit a rudimentary one. For instance, let us decide whether or not $p = 31418506212244678577$ is prime

```
> ifactor(31418506212244678577);
```

$$(7949)\,(7927)\,(7933)\,(7919)\,(7937)$$

```
> expand(%);
```

$$31418506212244678577$$

So p is not prime. The multipurpose function `expand` is doing just that, expanding the product, thereby allowing us to recover the original entry. This method of testing primes is inefficient, and it turns out that it is possible to decide whether or not an integer is prime without actually computing its prime decomposition.

Example 2.17. The function `isprime(x)` returns the value *true* if x is prime and the value *false* otherwise

```
> isprime(31418506212244678577);
```

$$false$$

```
> isprime(7949);
```

$$true$$

Strictly speaking, the above description of `isprime` is incorrect. This function runs a *probabilistic primality test,* meaning that when it returns the value *true,* the argument is 'very likely' to be a prime. In practice, not a single counterexample for which such test has failed is known, and if such counterexample exists, it has been conjectured to be hundreds of digits long. So you may trust `isprime` after all!

Example 2.18. *A trapdoor function.* If you run the commands `ifactor` and `expand` of the previous exercise, you will not fail to notice that factorization is a lot more time-consuming than multiplication. Multiplication is a *trapdoor function,* easy to do but very difficult to undo. This very interesting phenomenon is of great theoretical interest, and has applications in cryptography. To get a feeling for the behaviour of this trapdoor function, identify some large primes, multiply them together, and then attempt to undo the multiplication by means of `ifactor`. Monitor the computation time with the function `time` (see online documentation).

The Mersenne numbers*

The number $2^{1279} - 1$ computed in (2.7) is very interesting: it is prime, and so is 1279, the exponent of 2. Its primality was discovered in 1952 using a computer, and at that time it was the largest known prime. The numbers of the form

$$M_p = 2^p - 1 \qquad p \text{ prime}$$

are called *Mersenne numbers,* from the name of the 17th-century priest, Father Marin Mersenne, who first studied their properties. He originally thought that they were all primes, but this turned out not to be the case. For instance, we have already verified that M_{191} is not prime (see (2.12)), and indeed (as of December 2000) M_p has been shown to be prime only for the following values of p

$$2, 3, 5, 7, 13, 17, 19, 31, 61, 89, 107, 127, 521, 607, 1279, 2203, 2281,$$
$$3217, 4253, 4423, 9689, 9941, 11213, 19937, 21701, 23209, 44497,$$
$$86243, 110503, 132049, 216091, 756839, 859433, 1257787, 1398269,$$
$$2976221, 3021377, 6972593$$

$$(2.27)$$

It is not known whether there are infinitely many Mersenne primes. The importance of Mersenne numbers derives from the discovery, made by Lucas in 1878, of a very efficient algorithm for testing their primality. It was Lucas himself who found in this way the largest prime to be known before the computer age, namely

```
> 2^127-1;
```

$$170141183460469231731687303715884105727$$

The current world record corresponds to the largest prime in (2.27), that is, $2^{6972593} - 1$ with 2098960 digits.

By the time you read this page, this record may already have been broken: check the latest on Mersenne primes and other prime number records on the web sites

```
        http://www.mersenne.org
        http://www.utm.edu/research/primes/
```

The primality of small Mersenne numbers can be tested with the function `isprime`. Thus, with reference to (2.23) and (2.27)

```
> isprime(2^89-1);
```

$$true$$

```
> isprime(2^97-1);
```

$$false$$

Even though the algorithm used by `isprime` is quite sophisticated, this function does not implement Lucas' test, which works only for Mersenne primes. Maple has specialized software for Mersenne numbers — type `?mersenne` for details.

An advanced account of prime number records can be found in [4]. For the necessary background in arithmetic, see [5] or [2].

Exercises

Exercise 2.17.
(a) Determine the number of divisors of 20!
(b) Find all divisors of 472890989.
(c) Verify that $101^4 - 1$ has 5 prime divisors, without displaying them.

Exercise 2.18.
(a) By means of the function `prevprime`, determine the largest prime which is smaller than 10^8. (Obtain information about `prevprime` through the online help system.)
(b) Verify the result of part (a) directly, by testing for primality the integers less than 10^8, starting from $10^8 - 1$ and working your way backward. You must use the `isprime` function, but you should avoid testing those numbers that are divisible by 2, 3, or 5 (for these you do not need Maple!).

Exercise 2.19. There are 25 primes less than or equal to 100. Indeed, the 25th prime is 97, and the 26th prime is 101 — cf. (2.23).
(a) Verify the above statement with the function `ithprime`.
(b) Let $\pi(x)$ be the number of primes not greater than x (cf. 2.24). Use the function `ithprime` to determine $\pi(x)$ for $x = 1000, 2000, 3000, 4000, 5000$. (The function `nextprime` may also be useful.) This calculation is likely to require a certain amount of trial and error, but you must show only the minimal output that suffices to prove your result.
(c) Let $\Delta(x, n) = \pi(x) - \pi(x - n)$ be the number of primes that lie between $x - n$ and x ($x - n$ excluded). Tabulate the values of $\pi(x)$ and of $\Delta(x, 1000)$.

Exercise 2.20. Find out which is the largest Mersenne prime you can test for primality using `isprime`.

Exercise 2.21. Using the fundamental theorem of arithmetic, show that for any $x, y, z \in \mathbb{Z}$, we have

$$\gcd(x, y, z) = \gcd(\gcd(x, y), z) = \gcd(\gcd(x, z), y) = \gcd(\gcd(y, z), x).$$

This result says that the gcd of three integers can be computed by first computing the gcd of any two of them, and then the gcd of the latter and the third integer.

Exercise 2.22. Show that for any $m, n \in \mathbb{Z}$, not both zero, the following holds

$$\gcd(m, n) \cdot \operatorname{lcm}(m, n) = m\,n. \tag{2.28}$$

Exercise 2.23*. Find an integer greater than 10^{40} which has exactly 21 divisors. Such integer should be as small as possible: explain your strategy.

2.7 Standard library functions

In this chapter, we have encountered several functions supported by Maple. They are arithmetical functions such as `iquo`, `igcd`, and utility functions such as `subs`. In the table below, we list some common arithmetical functions. They belong to the so-called *standard library*, which is accessible automatically by a user at any time during a Maple session. These functions may be incorporated freely in any Maple expression.

ARITHMETICAL FUNCTIONS

function	symbolic name	example	value
General:			
absolute value	`abs(x)`	`abs(-41)`	41
factorial	`n!`	`9!`	362880
binomial coeff.	`binomial(x,y)`	`binomial(6,3)`	20
integer sq. root	`isqrt(x)`	`isqrt(19)`	4
maximum	`max(x,y,...)`	`max(-22,0)`	0
minimum	`min(x,y,...)`	`min(-2,0,-3)`	-3
No. of digits	`length(x)`	`length(1000!)`	2568
Divisibility:			
integer quotient	`iquo(x,y)`	`iquo(-11,5)`	-2
integer remainder	`irem(x,y)`	`irem(-11,5)`	-1
integer gcd	`igcd(x,y,...)`	`igcd(21,6,-33)`	3
integer lcm	`ilcm(x,y,...)`	`ilcm(21,12)`	84
Primes:			
prime factoriz.	`ifactor(x)`	`ifactor(24)`	$(2)^3 (3)$
primality test	`isprime(x)`	`isprime(7)`	*true*
*i*th prime	`ithprime(x)`	`ithprime(10)`	29
next prime	`nextprime(x)`	`nextprime(8)`	11
previous prime	`prevprime(x)`	`prevprime(11)`	7
Rationals:			
numerator	`numer(x)`	`numer(24/21)`	8
denominator	`denom(x)`	`denom(7)`	1
fractional part	`frac(x)`	`frac(-17/3)`	$-2/3$

Note that some Maple functions accept a single argument (e.g., `abs`, `ifactor`), and some accept two arguments (e.g., `binomial`, `iquo`). For other functions, the number of arguments is unspecified (e.g., `igcd`, `max`).

Example 2.19.   Because the evaluation of quotient and of remainder are closely related, one can compute both with a single call to either `iquo` or `irem`.   The function call `iquo(a,b,'r')` does not only compute the quotient of the division of a by b, but it also assigns the value of the remainder to the variable `r`. An analogous construction is possible for `irem`.

```
> iquo(27,7,'r'),r;
```
$$3, 6$$

```
> irem(27,7,'q'),q;
```
$$6, 3$$

Enclosing the third argument within right quotes is an essential precaution, because both functions require it to be an unassigned variable's name. (For an application, see section 5.1.)

Example 2.20.   When the argument is an integer, the multipurpose function `length` returns the number of its decimal digits, with the convention that the integer 0 has zero digits. If the argument is a symbol or a string, `length` will return the number of characters. This function accepts any data type as argument — see end of section 6.1.

Chapter 3

Sets and functions

In the previous chapter, we have come across a few important examples of *sets*, such as the set $\mathbb{N}$ of natural integers, the set $\mathbb{Z}$ of integers, and the set $\mathbb{Q}$ of rationals. In this chapter, we consider sets in greater generality. Then we define functions, which establish relations between sets.

3.1 Sets

A set is a collection (family, aggregate) of objects. We shall regard it as a primitive concept, without attempting to characterize it formally. The simplest way of defining a set is to list all its elements (whenever this is possible!), by enclosing them within curly brackets, separated by commas. For instance,

$$A = \{-1, 3, 5\}$$

is the set constituted by the three integers -1, 3, and 5. The order with which the elements of a set are listed is irrelevant, and therefore two sets are the same if they contain the same elements

$$A = \{-1, 5, 3\} = \{5, -1, 3\} = \{5, 3, -1\}, \quad \text{etc.}$$

If x belongs to A, we write $x \in A$, and $x \notin A$ otherwise. Thus, if $A = \{a, ab\}$, then $a \in A$ and $b \notin A$.

We shall agree that any repeated element is to be counted only once

$$A = \{7, 7, 0, 3, 7\} = \{7, 0, 3\}.$$

This is a commonly adopted convention: sets with repeated elements are often referred to as *multisets*.

It should be clear that the definition imposes no constraint on the type of objects that may constitute a set. Thus,

$$A = \{\odot, \ x^2 - 2 = 0, \ \text{Galapagos}\}$$

is a legitimate set, consisting of three elements — a symbol, an equation, and a name. Equally legitimate is a set containing another set as an element. For instance,

$$B = \{a, \{a, b\}\}$$

contains *two* elements, namely a and the set $\{a, b\}$. In particular,

$$a \in B \qquad b \notin B \qquad \{b, a\} \in B.$$

The *empty set* is the set that does not contain any element. This set is important in the formal development of the theory and is denoted by the symbol $\emptyset$. Formally

$$\emptyset = \{\}.$$

The number of elements of a set A — called the *cardinality* of A — is denoted by $|A|$.

$$|\{a, b, ab\}| = 3. \qquad |\{\{a, b\}\}| = 1 \qquad |\{\}| = 0.$$

Example 3.1. Some finite sets.

1. The set $D(n)$ of divisors of a positive integer n

$$\begin{aligned} D(120) &= \{1, 2, 3, 4, 5, 6, 8, 10, 12, 15, 20, 24, 30, 40, 60, 120\} \\ D(127) &= \{1, 127\}. \end{aligned} \tag{3.1}$$

2. The set $P(n)$ of all *partitions* of a positive integer n — the possible ways of representing n as a sum of positive integers, ignoring the order of the summands

$$P(5) = \{5,\ 4+1,\ 3+2,\ 3+1+1,\ 2+2+1,\ 2+1+1+1,\ 1+1+1+1+1\}.$$

3. The set $C(n)$ of all parenthesised expressions that evaluate to the product of n objects $a_1\, a_2 \cdots a_n$

$$\begin{aligned} C(4) = \{&a_1(a_2(a_3\, a_4)),\ a_1((a_2\, a_3)a_4),\ (a_1\, a_2)(a_3\, a_4), \\ &(a_1(a_2\, a_3))a_4,\ ((a_1\, a_2)a_3)\, a_4\}. \end{aligned}$$

Example 3.2. Some infinite sets.

1. The set of primes

$$\Pi = \{2, 3, 5, 7, 11, \dots\} \tag{3.2}$$

2. The set Π_n of primes that give remainder 1 when divided by n, for $n > 1$

$$\Pi_4 = \{5, 13, 17, 29, 37, \dots\}. \tag{3.3}$$

3. The set $n\mathbb{Z}$ of the multiples of a nonzero integer n

$$6\mathbb{Z} = \{\ldots, -12, -6, 0, 6, 12, \ldots\}. \tag{3.4}$$

Note that $-n\mathbb{Z} = n\mathbb{Z}$.

4. The set Φ_n of positive integers relatively prime to an integer $n > 1$

$$\Phi_6 = \{1, 5, 7, 11, 13, \ldots\}.$$

5. The set of squares in $\mathbb{Z}$

$$\{0, 1, 4, 9, 16, 25, \ldots\}.$$

6. The set $\mathbb{Z}[1/n]$ of rationals whose denominator is a power of a positive integer n. For example, the rational $127/128 = 127/2^7$ belongs to $\mathbb{Z}[1/2]$ while $128/127$ does not.

7. The set Ω_n of all words made of n letters, for $n > 0$

$$\Omega_2 = \{a, b, aa, bb, ab, ba, aaa, aab, \ldots\}. \tag{3.5}$$

8. The set of Pythagorean triples (see end of section 2.4)

$$\{(4, 3, 5), (12, 5, 13), (8, 15, 17), \ldots\}.$$

Example 3.3. Some sets whose status (finite vs. infinite) is not known.

1. The set $\mathcal{T}$ of *twin primes*, i.e., the set of pairs $(x, x + 2)$ where both x and $x + 2$ are prime

$$\mathcal{T} = \{(3, 5), (5, 7), (11, 13), (17, 19), (29, 31), (41, 43), \ldots\}. \tag{3.6}$$

The largest known twin primes (as of December 2000) are

$$665551035 \cdot 2^{80025} \pm 1$$

with 24099 digits.

2. The set of Mersenne numbers $M_p = 2^p - 1$, p prime, which are prime

$$\{M_2, M_3, M_5, M_7, M_{13}, \ldots\}.$$

3. The set of Mersenne numbers which are *not* prime

$$\{M_{11}, M_{23}, M_{29}, M_{37}, \ldots\}.$$

In the above examples, the sets were defined by listing the distinctive properties of their elements. We develop a notation for this type of definition. For instance, let A be the set of of non-negative powers of 2

$$A = \{1, 2, 4, 8, 16, 32, 64, 128, 256, \ldots\}.$$

We write

$$A = \{m \mid m = 2^k \text{ and } k \in \mathbb{N}\}.$$

The symbol before the vertical bar represents a generic element of the set. After the vertical bar, we list the properties that every element of the set must satisfy. The symbol before the bar may be replaced by an expression, resulting in greater conciseness

$$A = \{2^k \mid k \in \mathbb{N}\}.$$

If every element of A is an element of B, we say that A is *contained* in B or is a *subset* of B, and we write $A \subseteq B$. It then follows that for any set A we have $A \subseteq A$, and $\emptyset \subseteq A$. Every subset of A which is not A or $\emptyset$ — if it exists — is called a *proper* subset. If $A \subseteq B$ and $B \subseteq A$, then $A = B$. If $A \subseteq B$ but $A \neq B$, then we write $A \subset B$. (You should think of the relational operators $\subseteq$ and $\subset$ as being the set equivalent of $\leq$ and $<$, respectively.) The *power set* $\mathcal{P}(A)$ of a set A is the set of all its subsets, proper and improper

$$\mathcal{P}(\{a, b, c\}) = \{\{\}, \{a\}, \{b\}, \{c\}, \{a, b\}, \{a, c\}, \{b, c\}, \{a, b, c\}\}.$$

The relation $A \subseteq B$ is spelled out formally as

$$x \in A \quad \Longrightarrow \quad x \in B \tag{3.7}$$

where the symbol $\Longrightarrow$ reads 'implies' or — more mathematically — 'only if'. Expression (3.7) says that the property on the left-hand side of the arrow (the element x belongs to the set A) is true only if the property on the right-hand side (the element x belongs to the set B) is true. This suggests an alternative way of writing (3.7), namely,

$$x \notin A \quad \Longleftarrow \quad x \notin B$$

where $\Longleftarrow$ now reads 'if'. Note that reversing the implication requires negating the operator $\in$, that is, changing the logical value of both expressions (think about it).

Similarly, if $B \subseteq A$ we write

$$x \in A \quad \Longleftarrow \quad x \in B. \tag{3.8}$$

It should be clear that the relations $A \subseteq B$ and $B \subseteq A$ are not mutually exclusive. If (3.7) and (3.8) hold simultaneously, we have

$$x \in A \quad \Longleftrightarrow \quad x \in B$$

which says that x is in A 'if and only if' x is in B. The two statements are equivalent: they are either both true or both false. This is the formal definition of equality between sets $(A = B)$.

We now introduce operations between sets, defined in terms of the notion of membership to a set. These operations are *binary*, in that they involve two sets at a time, in much the same way as the arithmetical operations between numbers.

Union

The *union* of two sets A and B, denoted by $A \cup B$, is the set constituted by those elements that belong to *at least one* of the two sets A and B. In particular, the union includes the elements that belong to *both*.

$$x \in A \cup B \quad \Longleftrightarrow \quad x \in A \quad \text{or} \quad x \in B. \tag{3.9}$$

The conditions on the right-hand side, when taken individually, give two one-way implications

$$x \in A \cup B \quad \Longleftarrow \quad x \in A$$
$$x \in A \cup B \quad \Longleftarrow \quad x \in B.$$

Intersection

The *intersection* of two sets A and B, denoted by $A \cap B$, is the set constituted by those elements that belong to *both* A and B.

$$x \in A \cap B \quad \Longleftrightarrow \quad x \in A \quad \text{and} \quad x \in B. \tag{3.10}$$

We have two one-way implications

$$x \in A \cap B \quad \Longrightarrow \quad x \in A$$
$$x \in A \cap B \quad \Longrightarrow \quad x \in B.$$

Difference

The difference of B from A, denoted by $A \setminus B$, is the set of the elements of A that *do not* belong to B.

$$x \in A \setminus B \quad \Longleftrightarrow \quad x \in A \quad \text{and} \quad x \notin B. \tag{3.11}$$

We have two one-way implications

$$x \in A \setminus B \quad \Longrightarrow \quad x \in A$$
$$x \in A \setminus B \quad \Longrightarrow \quad x \notin B.$$

Example 3.4. Let $A = \{a, b, c\}$ and $B = \{a, c, d, e\}$. We have $A \cup B = \{a, b, c, d, e\}$, $A \cap B = \{a, c\}$, $A \setminus B = \{b\}$, and $B \setminus A = \{d, e\}$.

Example 3.5. Let $A \subseteq B$. Then $A \cup B = B$, $A \cap B = A$, and $A \setminus B = \emptyset$.

Example 3.6. Let $\mathbb{N}_e$ be the set of even natural numbers and $\mathbb{N}_o$ that of odd natural numbers. Then $\mathbb{N}_e \cup \mathbb{N}_o = \mathbb{N}$, $\mathbb{N}_e \cap \mathbb{N}_o = \emptyset$, $\mathbb{N}_e \setminus \mathbb{N}_o = \mathbb{N}_e$, $\mathbb{N}_o \setminus \mathbb{N}_e = \mathbb{N}_o$.

Example 3.7. With reference to (3.2) and (3.3), we have that

$$\Pi = \Pi_2 \cup \{2\}.$$

Example 3.8. For any nonzero integers m and n we consider the sets $m\mathbb{Z}$ and $n\mathbb{Z}$, defined in (3.4). Every element of $n\mathbb{Z} \cap m\mathbb{Z}$ is a multiple of both n and m, and every multiple of n and m is an element of $n\mathbb{Z} \cap m\mathbb{Z}$. Therefore, this set contains the least common multiple of n and m. But it cannot contain any smaller positive integer, from the definition of least common multiple. So we have

$$m\mathbb{Z} \cap n\mathbb{Z} = \mathrm{lcm}(m,n)\mathbb{Z}.$$

Example 3.9. Consider the set $D(n)$ of the divisors of n (see (3.1)). Then the common divisors of m and n constitute the set $D(m) \cap D(n)$. Because every divisor of m and n is a divisor of their greatest common divisor, we have

$$D(\gcd(m,n)) = D(m) \cap D(n).$$

Exercises

Exercise 3.1. Prove that for any two sets A and B, the following holds

$$A \cap B \subseteq A \qquad A \cap B \subseteq B \qquad A \cap B \subseteq A \cup B.$$

Exercise 3.2. Let S and C be the sets of squares and cubes of natural numbers, respectively.

$$S = \{0,1,4,9,16,\ldots\} \qquad C = \{0,1,8,27,64,\ldots\}$$

Characterize $S \cap C$ in terms of the prime factorization of its elements.

Exercise 3.3. Show that the set of primes p such that $p+2$ and $p-2$ are also primes has only one element. (*Hint:* if p is prime, then one of $p+2$ or $p-2$ is divisible by 3.)

Exercise 3.4*. Let A, B, and C be arbitrary sets. Prove that (i) $A \cup (B \cup C) = (A \cup B) \cup C$ (associative law for unions); (ii) $A \cap (B \cap C) = (A \cap B) \cap C$ (associative law for intersections); (iii) $A \cap (B \cup C) = (A \cap B) \cup (A \cap C)$ (distributive law).

3.2   Sets with Maple

Maple supports the *set* data type, with straightforward syntax. The following assignment statement defines a set of 3 integers

```
> T:={1,3,-4};
```

$$T := \{1, 3, -4\}$$

```
> whattype(%);
```

$$set$$

A set is an example of a *composite data type,* which can consist of an arbitrary number of operands. Maple eliminates repeated elements in a set

```
> {7,7,0,3,7};
```

$$\{7, 0, 3\}$$

and may rearrange the order of the elements

```
> {-4,3 1};
```

$$\{1, 3, -4\}$$

To check whether or not a given element belongs to a set, we use the standard library function `member`, which returns a value of the *logical* type

```
> U:={a,{a,b}}:
> member(a,U),member(b,U),member({b,a},U);
```

$$true, \; false, \; true$$

The empty set $\emptyset$ is represented in Maple by the curly brackets alone

```
> empty:={};
```

$$empty := \{\}$$

To count the number of elements of a set, we use the function `nops` (number of operands)

```
> nops(U),nops(empty);
```

$$2, 0$$

Set operators

Maple supplies three *set operators,* namely `union`, `intersect`, and `minus`, for union, intersection, and difference between sets, respectively (see (3.9), (3.10), and (3.11)). Their straightforward syntax is illustrated in the following examples.

```
> A:={1,{1,2}};
```

$$A := \{1, \{1, 2\}\}$$

```
> B:={1,{1,3}};
```

$$B := \{1, \{1, 3\}\}$$

```
> A union B;
```

$$\{1, \{1, 2\}, \{1, 3\}\}$$

```
> A intersect B;
```

$$\{1\}$$

```
> A minus B;
```

$$\{\{1, 2\}\}$$

```
> B minus A;
```

$$\{\{1, 3\}\}$$

Logical operators

Sets are often defined through set operations. For instance, let a and b be rationals such that $a < b$. The rational interval $[a, b)$ containing the left endpoint a but not the right endpoint b is the intersection of two sets A and B, given by

$$A = \{x \in \mathbb{Q} \mid x \geq a\} \qquad B = \{x \in \mathbb{Q} \mid x < b\}.$$

Now, from the definition (3.10) of the intersection of two sets, we may say that x belongs to $[a, b) = A \cap B$, when the proposition $x \in A$ is *true* and the proposition $x \in B$ is *true*. Accordingly, we shall test the condition $x \in [a, b)$ with Maple through the truth or falsehood of the *logical expression*

```
> x >= a and x < b;
```

The union of two sets would require a similar construct, with the term 'and' replaced by 'or', whereas the difference between A and B would read $x \in A$ and (not $x \in B$).

The binary *logical operators* **and** and **or**, and the unary logical operator **not**, relate expressions whose value is of the logical (Boolean) type. These operators act on the set $\mathcal{B} = \{true, false\}$ of the two Boolean constants. (Maple actually has a three-way logic, involving the additional Boolean constant FAIL, but we shall ignore it for the time being.) They should be thought as an analogue of the binary arithmetical operators $+$ and $\times$, and the unary operator $-$ (sign change, carrying x to $-x$), respectively, which combine, for instance, elements of $\mathbb{Z}$ and $\mathbb{Q}$.

The action of the binary operators and and or and of the unary operator not is defined by listing all possibilities:

LOGICAL OPERATORS

x	y	x and y	x or y	not x
false	*false*	*false*	*false*	*true*
true	*false*	*false*	*true*	*false*
false	*true*	*false*	*true*	*true*
true	*true*	*true*	*true*	*false*

Thus, x and y is equal to *true* only if both operands are equal to *true*, while for x or y to be equal to *true*, only one of the operands needs to have the value *true*. The following properties of logical operators can be verified directly from the entries of the above table

$$(i) \qquad x \text{ and } y = y \text{ and } x$$
$$(ii) \qquad x \text{ or } y = y \text{ or } x$$
$$(iii) \qquad \text{not } (x \text{ and } y) = (\text{not } x) \text{ or } (\text{not } y)$$
$$(iv) \qquad \text{not } (x \text{ or } y) = (\text{not } x) \text{ and } (\text{not } y)$$

The first two properties state the *commutativity* of the operators and and or, which in this respect behave just like $+$ and $\times$.

Maple implements logical operators and expressions in a straightforward manner. If there are several operators in a logical expression, they are evaluated in the following order, given by: first not, then and, then or. Parentheses may be used to alter the order of evaluation. Expressions containing logical operators are identified by Maple as being of the logical type, and evaluated automatically. When this is the case, the use of `evalb` is no longer required.

In the following examples we combine logical operators with relational expressions:

Example 3.10. We verify property (iii) above, for the values $x = true$, $y = false$

```
> not (true and false) = (not true) or (not false);
```
$$true$$

Example 3.11. Let $A = \{1, 2, 3\}$ and $B = \{2, 3, 4\}$. We verify that $4 \in B \setminus A$ in two ways, using the two equivalent definitions (3.11)

```
> A:={1,2,3}:   B:={2,3,4}:
> member(4, B minus A), member(4,B) and not member(4,A);
```
$$true, true$$

Example 3.12. Verify that 83 divides $3^{41} - 1$, but it does not divide $2^{41} - 1$.

```
> not irem(2^41-1,83)=0 and irem(3^41-1,83)=0;
```
$$true$$

Even if the relational operator = is involved, the presence of the logical operators makes the Boolean status of the above expression unequivocal, so that Maple evaluates it automatically without the need of calling `evalb`.

Example 3.13. Verify that the integer $n = 1004^2 + 1$ is a prime of the form $7k + 3$ or $7k + 4$, for some integer k. We have to test n for primality and check that n gives remainder 3 or 4 when divided by 7.

```
> 1004^2 + 1:
> isprime(%) and (irem(%,7)=3 or irem(%,7)=4);
```
$$true$$

The last expression requires two evaluations of `irem(%,7)`. One of the two can be avoided as follows

```
> isprime(%) and member(irem(%,7),{3,4});
```

Exercises

Exercise 3.5. Without displaying any digit, verify that 11396333 is divisible by 43 or 47, but not by both.

3.3 Functions

Let A and B be two sets. By a *function* f defined on A with values in B we mean a rule which associates to every element x of A one — and only one — element of B, denoted by $f(x)$. We write

$$f : A \to B \qquad f := x \mapsto f(x). \qquad (3.12)$$

The set A is called the *domain* of f, and the set B the *co-domain* (or *range*).

The clause that $f(x)$ be unique is crucial: this is what makes functions interesting. Also, a function must be defined for *all* elements of the domain A. If this is not the case, then A cannot be called the domain of the function. In other words, the domain is always chosen as economically as possible: it has no redundant elements.

As x varies in the domain A, $f(x)$ varies in the co-domain B, and the set of elements of B that can be obtained in this way is called the *image* of A, denoted by $f(A)$. Because every element of $f(A)$ is in B, we have $f(A) \subseteq B$, but the definition of a function does not require that $f(A)$ be equal to B, that is, that the image must coincide with the co-domain. So, unlike the domain, the co-domain of a function can be — and often is — chosen with redundant elements.

If the image coincides with the co-domain, we say that f is *surjective* (or *onto*). Then, for every element y of B there exists at least one element x of A such that $f(x) = y$. It immediately follows that any function f can be turned into a surjective function if we identify co-domain and image: $f : A \to f(A)$.

If distinct elements of A always have distinct images in B, then f is said to be *injective* (or *one-to-one*). Thus, a function is injective if $f(x) = f(y)$ implies $x = y$, for all x and y in the domain. If the domain of f is finite, then f is injective precisely when domain and image have the same cardinality (think about it).

A function which is both injective and surjective is said to be *bijective*. A bijective function establishes a strong correspondence between domain and co-domain.

Example 3.14. On any set A, we define the *identity function,* given by

$$I_A : A \to A \qquad I_A := x \mapsto x.$$

The identity function is bijective, and its domain, co-domain, and image all coincide with A.

Example 3.15. Let

$$A = \{-2, -1, 0, 1, 2\} \qquad B = \{-4, -3, -2, -1, 0, 1, 2, 3, 4\}$$

and

$$f : A \to B \qquad f := x \mapsto 2x.$$

Then $f(A) = \{-4, -2, 0, 2, 4\}$, so that the image is distinct from the co-domain and f is not surjective. For instance, $-3 \in B$, but there is no element x of A such that $f(x) = -3$. Because the domain of f is finite, and the image has the same cardinality as the domain, we conclude that f is injective without further check.

Example 3.16. The function $g := x \mapsto x^2$, with domain and co-domain as in the previous example, is neither injective nor surjective. Indeed, $|A| = 5$ and $|g(A)| = 3$, so g is not injective (for instance, -1 and 1 are distinct elements of A but $g(-1) = g(1) = 1$ is the same element of B). Likewise, $-1 \notin g(A)$, so g is not surjective.

Example 3.17. Let

$$A = \{0, 1, 2, 3, 4\} \qquad B = \{0, 2, 4, 6, 8\}$$

and

$$f : A \to B \qquad x \mapsto \operatorname{rem}(6x, 10).$$

Thus, $f(0) = 0$, $f(1) = 6$, $f(2) = 2$, $f(3) = 8$, and $f(4) = (4)$. This function is bijective.

Example 3.18. Consider the following function

$$f : \mathbb{Z} \setminus \{0\} \to \mathbb{Q} \qquad f := x \mapsto \frac{x+1}{x}. \qquad (3.13)$$

Division by zero requires that the integer $x = 0$ be excluded from the domain of f, for otherwise we would have to set $f_0 = \infty$, but this value is not a rational number. Alternatively, one could adjoin the point $x = \infty$ to the co-domain of f, and extend the definition of f to $x = 0$ as follows:

$$f : \mathbb{Z} \to \bar{\mathbb{Q}} \qquad f := x \mapsto \begin{cases} \frac{x+1}{x} & \text{if } x \neq 0, \\ \infty & \text{if } x = 0 \end{cases}$$

where $\bar{\mathbb{Q}} = \mathbb{Q} \cup \{\infty\}$. With either definition, the function f is injective. We prove it for the case (3.13). If $f(x) = f(y)$, then $(x+1)/x = (y+1)/y$, and because x and y are nonzero, we have $xy + y = xy + x$, whence $x = y$. On the other hand, f is not surjective. For instance, the rational 3 does not belong to the image $f(\mathbb{Z})$, because the equation $f(x) = 3$ yields $x = 1/2$, which is not in the domain $\mathbb{Z}$.

Example 3.19. Let S be the set of squares of elements of $\mathbb{N}$

$$S = \{0^2, 1^2, 2^2, 3^2, 4^2, 5^2, \ldots\}.$$

We define the *square root* function as follows

$$\sqrt{} : S \to \mathbb{N} \qquad \sqrt{} := x \mapsto y, \text{ where } y^2 = x.$$

The integer $y = \sqrt{x}$ is uniquely defined by x, because we have taken the precaution of excluding the negative integers from the co-domain. Therefore, if one keeps the above definition, the co-domain of the square root cannot be extended to $\mathbb{Z}$, because to every x in S there would correspond *two* elements of $\mathbb{Z}$, contradicting the defining property of a function.

Example 3.20. Let Π be the set of primes. We define

$$\tau : \Pi \to \Pi \cup \{0\} \qquad p \mapsto \begin{cases} p & \text{if } p + 2 \in \Pi \\ 0 & \text{otherwise.} \end{cases}$$

The function τ is neither injective nor surjective, which can be seen, for instance, from the fact that $\tau(2) = \tau(7) = 0$: two elements of the domain have the same image, and neither of them belongs to $\tau(\Pi)$. The set $\tau(\Pi)$ consists of 0 and a representative for each pair of *twin primes*; see (3.6).

3.4   User-defined functions

In Maple, it is possible to define new functions, to be used alongside the library functions. They are the *user-defined* functions. The simplest such construction makes use of the *arrow operator* ->. For example, the function defined in (3.13) is constructed in Maple as follows

```
> f:=x->(x+1)/x;
```

$$f := x \to \frac{x+1}{x}$$

This is a new data type

```
> whattype(%);
```

$$procedure$$

Once a new function is defined, it can be used like any other library function, by substituting any valid expression for its argument. In our example, valid expressions are arithmetical and algebraic expressions

```
> f(2/3),f(joe),f(a^2-1);
```

$$\frac{5}{2}, \frac{joe+1}{joe}, \frac{a^2}{a^2-1}$$

We make some syntactical remarks.

The name of the function's argument (the variable appearing immediately on the left of the arrow operator) is invisible from the outside, and in particular, it is unrelated to any variable with the same name that could have been previously defined. Study the following example carefully:

```
> g:=n->2*n:
> h:=whatever->2*whatever:
> whatever:=3:
> g(2),h(2),g(whatever),h(whatever),g(n),h(n);
```

$$4, 4, 6, 6, 2n, 2n$$

Common mistakes originate from assignments such as

```
> wrong(x):=2*x+1:
```

This statement does *not* define a Maple procedure, in spite of the fact that it is syntactically legal. Thus,

```
> wrong(x),wrong(2),wrong(1/y);
```

$$2x+1, wrong(2), wrong\left(\frac{1}{y}\right)$$

One sees that Maple knows what is wrong(x), but nothing else (this feature, which allows a function to be defined at individual values of the argument, is useful in the so-called recursive definitions).

In the above construct, the value assigned to the variable f is a *procedure definition:* $x \to (x+1)/x$. However, for reasons of efficiency of computation, the function's symbolic name is not evaluated automatically (the function's name is a *pointer* to the place in memory where its value is stored)

```
> f,whattype(f);
```

$$f, \; symbol$$

To force evaluation, one must use the intrinsic function `eval`

```
> eval(f),whattype(eval(f));
```

$$x \to \frac{x+1}{x}, \; procedure$$

Results analogous to a procedure definition can also be obtained via substitutions in an expression, although this is often clumsier and less transparent.

```
> f:=(x+1)/x;
```

$$f := \frac{x+1}{x}$$

```
> subs(x=2/3,f),subs(x=joe,f),subs(x=a^2-1,f);
```

$$\frac{5}{2}, \; \frac{joe+1}{joe}, \; \frac{a^2}{a^2-1}$$

Finally, a fine point of mathematical notation is noted. The arrow operator `->` corresponds to the mathematical symbol $\mapsto$ for function definition — see equation (3.12). However, Maple displays it as $\to$, which has a different meaning, namely that of specifying domain and co-domain.

Characteristic functions and the `if`-structure

Let A be a set, and let C be a subset of A. The function

$$\chi_C : A \to \{0,1\} \qquad\qquad x \mapsto \begin{cases} 1 & \text{if } x \in C \\ 0 & \text{if } x \notin C \end{cases} \qquad (3.14)$$

is called the *characteristic function* of C in A. A variant of the above construction is the *Boolean characteristic function*, given by

$$\chi_C : A \to \{true, false\} \qquad\qquad x \mapsto \begin{cases} true & \text{if } x \in C \\ false & \text{if } x \notin C. \end{cases} \qquad (3.15)$$

Example 3.21. For a non-empty set A the following holds

$$\chi_\emptyset(A) = \{0\} \qquad\qquad \chi_A(A) = \{1\}.$$

So if C is an *improper* subset of A (i.e., $C = A$ or $C = \emptyset$), χ fails to be surjective. On the other hand, let C be a *proper* subset of A. Because C is not empty, it has at least one element, whence $\chi_C(A)$ contains 1. But since $C \neq A$, there exists at least one element of A which is not in C, whence $\chi_C(A)$ contains 0. Therefore, the characteristic function χ_C is surjective if and only if C is a *proper* subset of A.

The evaluation of the characteristic function (3.14) or (3.15) requires a process of decision-making: we have to decide whether or not the element x

belongs to the set C, and then assign the value to the function χ accordingly. The Boolean case is simpler, since the values *true* or *false* can be obtained by evaluating a Boolean expression. If the set C is given explicitly, then $\chi_C(x)$ will be represented in Maple by the expression `member(x,C)`. If C is defined by a certain property, then one will translate the latter into a Maple *logical expression*, and then adopt a construct of the type

```
> chi:=x->logical expression;
```

which may or may not require the use of `evalb`.

Example 3.22. Let n be an integer. We construct the function `even(n)` whose value is *true* if n is even, and *false* if n is odd. This is the Boolean characteristic function of $2\mathbb{Z}$ in $\mathbb{Z}$.

```
> even:=n->evalb(irem(n,2)=0):
```

Example 3.23. We consider the interval $[0, 1)$ in $\mathbb{Q}$ consisting of all rationals between 0 and 1 (the left endpoint 0 is included, the right endpoint 1 is not). Its Boolean characteristic function is given by

```
> interval:=x->x >= 0 and x < 1:
```

Note that `evalb` is unnecessary here, since the presence of the logical operator `and` makes the logical status of the expression unambiguous.

In the non-Boolean case (3.14), the decision-making is implemented in Maple by the `if`-structure. Its simplest form is the following:

```
> if logical expression then expression 1 else expression 2 fi:
```

If the value of *logical expression* is *true* then *expression 1* is executed; if it is *false*, then *expression 2* is executed.

Two syntactical remarks. First, the evaluation of the logical expression is done automatically; that is, there is no need to use `evalb`. Second, the statements need not be followed by a terminator (colon or semicolon). This is because an `if`-statement is logically equivalent to a single statement. We shall deal with the `if`-structure in greater generality in chapter 9. Here we are mainly interested in its applications to the construction of characteristic functions.

Example 3.24. Consider the characteristic function of the subset $\{0\}$ of $\mathbb{Z}$ (cf. (3.14)).

$$\chi : n \mapsto \begin{cases} 1 & \text{if } n = 0 \\ 0 & \text{if } n \neq 0. \end{cases} \tag{3.16}$$

The Maple implementation is straightforward:

```
> chi:=n->if n=0 then 1 else 0 fi:
```

The Boolean version of the same function is even simpler, as it does not require the `if`-structure

```
> chi:=n->evalb(n=0);
```

Example 3.25.   The Boolean characteristic function of the set of primes is given by `isprime`. Its non-Boolean version is then constructed as

```
> chi:=n->if isprime(n) then 1 else 0 fi:
```

Example 3.26.   We consider the characteristic function of $\mathbb{Z}$ in $\mathbb{Q}$

$$\chi : \mathbb{Q} \to \{0,1\} \qquad \chi : x \mapsto \begin{cases} 1 & \text{if } x \in \mathbb{Z} \\ 0 & \text{if } x \notin \mathbb{Z}. \end{cases}$$

We note that a rational x is an integer precisely when its denominator is equal to 1. Thus,

```
> chi:=x->if denom(x)=1 then 1 else 0 fi:
```

Example 3.27.   To illustrate a more general usage of the if-structure, we construct the function `nint(x)`, which returns the integer nearest the positive rational x. If the fractional part of $x = a/b$ does not exceed $1/2$, the nearest integer is the quotient of the division of a by b; otherwise, it is the same quantity plus one.

```
> nint:=x->if 2*irem(numer(x),denom(x))<=denom(x) then
>               iquo(numer(x),denom(x))
>           else
>               iquo(numer(x),denom(x))+1
>           fi:
```

To improve readability, we have subdivided the expression over several lines of input. The above version of `nint` has two shortcomings: first, it gives the incorrect answer when the argument is negative, and second, it is inefficient in that the functions `irem` and `iquo` perform essentially the same calculation (see remarks at the end of section 2.7). To solve these problems we shall require a more general form of procedure, discussed in section 9.4.

Functions of several variables

The arrow operator construction can be generalized to the case of functions of several variables. The formal aspect of this construction will be dealt with in chapter 9. Here it suffices to note that a function of several variables is a rule that associates to a sequence $(x_1, x_2, \ldots, x_n)$ of n elements of the sets $A_1, \ldots, A_n$ a unique element of another set, which we denote by $f(x_1, x_2, \ldots, x_n)$. In Maple, the sequence of variables must be enclosed within round brackets.

Example 3.28.   *A function depending on a parameter.* Let a be an integer, and let us consider the function

$$f_a : \mathbb{Z} \to \mathbb{Z} \qquad f_a := x \mapsto ax.$$

Every value of a corresponds to a different function. If $a = 0$ the image of f_a is the set $\{0\}$, and f_a is neither injective nor surjective. If $a \neq 0$, the image of $\mathbb{Z}$ under f_a is the set of multiples of a, that is, the set of all integers that are divisible by a. If $|a| = 1$, then $f_a(\mathbb{Z}) = \mathbb{Z}$, and f_a is bijective. For $a = 1$, f_a is the identity function on $\mathbb{Z}$. If $|a| > 1$, f_a is injective but not surjective. Convince yourself that for any value of a, $f_a(\mathbb{Z}) = f_{-a}(\mathbb{Z})$. To represent this function with Maple, we think of it as a function of the two variables a and x

```
> f:=(a,x)->a*x;
```

$$f := (a, x) \to a\,x$$

```
> f(3,k),f(3,3);
```

$$3k, 9$$

Example 3.29. We construct the characteristic function of the set $d\mathbb{Z}$ of the integer multiples of a nonzero integer d, defined in (3.4). This is a function of the two variables x and d, where d is regarded as given, while x assumes any integer value. Using the fact that `irem(x,d)` is zero precisely when n is a multiple of d, we have

```
> Multd:=(x,d)->if irem(x,d)=0 then 1 else 0 fi:
```

Example 3.30. We construct the characteristic function `Divs(x,n)` of the set of divisors (positive or negative) of a given nonzero integer n in $\mathbb{Z}$. We begin with a preliminary definition:

```
> Divs:=(x,n)->if irem(n,x)=0 then 1 else 0 fi;
```

As for `Multd`, the construction is based on `irem`, with the role of the variables interchanged. This time, the first variable n is fixed, while the second one is allowed to vary in the domain $1 \le |x| \le n$. The following alternative definition makes the relationship between the two functions explicit.

```
> Divs:=(x,n)->Multd(n,x):
```

If the first argument of `Divs` is equal to zero, we get an error termination, because the second argument of `irem` cannot be zero. On the other hand, it would be desirable to allow zero as a legal first argument, in which case `Divs` should return the value zero. This can be achieved as follows

```
> Divs:=(x,n)->if not x=0 and irem(n,x)=0 then 1 else 0 fi:
```

The logical expression defining `Divs` is evaluated left to right, and if $x = 0$ the relational expression `not x=0` evaluates to *false*. Maple knows that the entire logical expression is then *false*, so the relational expression `irem(n,x)=0` is not evaluated at all. Note that the logically equivalent construct

```
> Divs:=(x,n)->if irem(n,x)=0 and not x=0 then 1 else 0 fi:
```

would not work, because in this case, Maple evaluates `irem(n,x)` before checking that x is nonzero.

Mapping all elements of a set

The image $f(A)$ of a set A under a function f was defined as the set obtained applying f to all elements of A. The image of a set can be constructed in Maple with the standard library function `map`.

In the following example, we construct the image of the set $\{-2, -1, 0, 1, 2\}$ under the function $f : x \mapsto x^2$

```
> A:={-2,-1,0,1,2}:
> f:=x->x^2:
> map(f,A);
```

$$\{0, 1, 4\}$$

The first argument of `map` is an expression of the *procedure* type (a function definition), while the second is a *set*. (In chapter 5 we shall see that `map` accepts any expression as a second argument, not just a set.) Because the *value* of the variable `f` is `x->x^2`, such expression can be used explicitly, without defining the function separately

```
> map(x->x^2,A);
```

Example 3.31. Let $A = \{0, 1, 2, 3, 4, 5\}$. We show with Maple that the mapping $f : x \mapsto \mathrm{rem}(x + 5, 6)$ of A to itself is bijective. Because domain and co-domain are the same, it suffices to check that $A = f(A)$.

```
> {0,1,2,3,4,5};
> evalb(map(x->irem(x+5,6),%)=%);
```

$$true$$

Example 3.32. Let A be a set of integers. We construct a function which returns *true* if A contains a prime, and *false* otherwise.

```
> ContainsPrimes:=A->member(true,map(isprime,A)):
```

The expression `map(isprime,A)` is the image of A under the function `isprime`. If there is a prime in A, then such image contains the element *true*, which is the test performed by the function `member`.

If the first argument of `map` is a function of several variables, then the value of all variables beyond the first one must be supplied in the form of a sequence of optional arguments. We illustrate this construction in the following examples.

Example 3.33. Let A be a finite set of integers, and let n be any integer. Then the value of the expression `map(igcd,A,n)` is the set constituted by the greatest common divisors of n and each element of A.

Example 3.34. Let A be a set of integers, and let d be a nonzero integer. We construct a function mapping A to the set of rationals obtained by dividing each element of A by d.

```
> DivBy:=(A,d)->map((x,y)->x/y,A,d):
```

Exercises

Exercise 3.6.   Construct a user-defined function for the function $h(a) = a^2 + a - 1$. Use this function to compute the following expressions

$$5^2 + 5 - 1 \qquad\qquad 3^2 - 3 - 1 \qquad\qquad -b^4 + b^2 + 1$$

$$\frac{1}{(x+1)^2} + \frac{1}{x+1} - 1 \qquad\qquad \frac{1}{Z^6 + Z^3} \qquad\qquad (11^4 - 1)^2 - 11^4.$$

Exercise 3.7.   Let $A = \{1, 2, 3\}$ and $B = \{0, 1\}$.

(a) Determine the number of distinct surjective functions $f : A \to B$.

(b) Construct three distinct user-defined surjective functions $f : A \to B$ [*Hint:* use `irem`, `iquo`, `abs`, etc.].

(c) Using one of the functions f constructed above and `map`, verify directly that f is surjective.

Exercise 3.8.   Construct user-defined functions for the following characteristic functions

(a) $f(x) = \begin{cases} 1 & \text{if } x \text{ is negative} \\ 0 & \text{otherwise} \end{cases}$

   Use f to decide whether or not $500^2 - 89 \cdot 53^2$ is negative.

(b) $f(x) = \begin{cases} 1 & \text{if } x \text{ is a multiple of 7} \\ 0 & \text{otherwise} \end{cases}$

   Use f to decide whether or not $100^2 + 98^2$ is divisible by 7.

(c) $f(x) = \begin{cases} 1 & \text{if } x \text{ and } x + 2 \text{ are prime} \\ 0 & \text{otherwise} \end{cases}$

   Use f to decide whether or not $p_{105} - p_{104} = 2$, where p_k is the kth prime. (Think about it.)

(d) $f(x, y) = \begin{cases} 1 & \text{if } x \text{ and } y \text{ are relatively prime} \\ 0 & \text{otherwise} \end{cases}$

   Use f to decide whether or not the prime 2999 divides $9! - 1$.

Exercise 3.9.   Construct the Boolean characteristic function of the following sets. Use `evalb` only if necessary.

(a) The set of even non-negative integers.

(b) The set of rationals whose numerator is odd.

(c) The set of rationals whose denominator is composite.

(d) The set of primes which are twice a prime plus one.

(e) The set of integers which are divisible by 5 or by 7, but not by both.

(f) The set of positive integers i for which the ith prime is twice a prime plus one; hence compute the intersection between such set and the set $\{101, 102, 103, 104, 105\}$.

Exercise 3.10. Let n be a positive integer. Construct a function `nearp(n)` whose value is the prime number nearest n.

Exercise 3.11. Let r and s be positive rational numbers. Construct a function `K(r,s)` whose value is *true* if there is an integer between r and s, and *false* otherwise. (You may assume that r and s are not integers.)

Chapter 4

Sequences

Sequences are functions defined over the integers. The ordering property of the integers is what makes sequences special.

4.1 Basics

When the domain of a function is the set of positive integers $\mathbb{N}^+$, one speaks of a *sequence*. Thus, a sequence f is a rule which associates to every positive integer n a unique element $f(n)$ of a given set B. For instance, let $B = \mathbb{Z}$, and let us consider the function f which associates to every natural number n the value of the expression $2n - n^2$. We write

$$f : \mathbb{N}^+ \to \mathbb{Z} \qquad f := n \mapsto 2n - n^2$$

or, more compactly

$$f(n) = 2n - n^2 \qquad n \geq 1. \tag{4.1}$$

The elements of this sequence are the integers $f(1), f(2), f(3), \ldots = 1, 0, -3, \ldots$. When a function is a sequence, it is customary to write f_n in place of $f(n)$. So we write out the elements of the sequence as

$$f_1, f_2, f_3, \ldots, f_k, \ldots$$

and the entire sequence as $\{f_k\}$, or, more pedantically, $\{f_k\}_{k=1}^{\infty}$. In the above notation, f is a function, k is an element of the domain of f, f_k is an element of the co-domain of f, and $\{f_k\}$ is the image (think about it).

The condition that the domain of a sequence be $\mathbb{N}^+$ may be relaxed. For instance, it may be convenient to restrict the domain to a proper subset of the positive integers, or to extend it to the natural numbers $\mathbb{N}$, or even to the whole of $\mathbb{Z}$. In the latter case, we speak of a *doubly-infinite sequence*

$$\ldots, a_{-2}, a_{-1}, a_0, a_1, a_2, \ldots$$

For instance, the sequence of odd integers is doubly-infinite.

Example 4.1.   Let us consider the following sequence of rational numbers

$$a : n \mapsto \frac{1}{n^2 - 3n + 2}. \tag{4.2}$$

Factoring the denominator $n^2 - 3n + 2 = (n - 1)(n - 2)$, we see that (4.2) is undefined for $n = 1, 2$ when the denominator vanishes and a_n becomes infinite. Accordingly, we restrict the domain of a to the subset of $\mathbb{N}^+$ given by $n \geq 3$, which is $\mathbb{N}^+ \setminus \{1, 2\}$. Alternatively, we may perform the change of variables $n = m - 2$. Then the inequality $n \geq 3$ becomes $m \geq 1$, and (4.2) is transformed into a new sequence defined over the whole of $\mathbb{N}^+$

$$b(m) = \frac{1}{(m - 2)^2 - 3(m - 2) + 2} = \frac{1}{m^2 - 7m + 12} \qquad m \geq 1.$$

It should be clear that $b_k = a_{k+2}$ for all $k \geq 1$.

It must be stressed that the co-domain B that contains the elements of a sequence, as well as the nature of the rule associating an element of B to every positive integer, is completely arbitrary. In particular, one should not assume that the computation of $f(n)$ merely requires substituting the value of n in some expression

$$f := n \mapsto \text{ an expression depending on } n \qquad n \geq 1. \tag{4.3}$$

This is just the simplest possibility, and the sequences constructed in this way form a small class among all sequences.

Example 4.2.   We consider the sequence generated by the Boolean characteristic function of the set of primes in $\mathbb{N}^+$:

$$p := n \mapsto \begin{cases} true & \text{if } n \text{ is prime} \\ false & \text{if } n \text{ is not prime} \end{cases} \qquad n \geq 1. \tag{4.4}$$

The elements of such a sequence belong to the set $\{true, false\}$: the first few of them are

$$false, true, true, false, true, false, true, false, false, false, \ldots$$

Even though the function p is represented explicitly by the function `isprime`, the process that allows Maple to determine whether $p(n)$ is $true$ or $false$ is complicated. In spite of some 2000 years of effort, nobody has yet succeeded in reducing this algorithm to the evaluation of a simple 'formula'. If you test `isprime` on large values of n, you will notice how much Maple struggles to give you the answer.

4.2 Sequences with Maple

A sequence defined explicitly (see (4.3)) can be represented in Maple by any function taking integers as arguments. Such function may be a Maple library function, or one defined by the user. In either case, there is a straightforward way of generating its elements, with the Maple intrinsic function seq. The following command generates the first 10 prime numbers, by making use of the standard library function ithprime.

> seq(ithprime(k),k=1..10);

$$2,\ 3,\ 5,\ 7,\ 11,\ 13,\ 17,\ 19,\ 23,\ 29$$

The intrinsic function seq has two arguments. The first is the Maple expression ithprime(k), while the second specifies the variable k, as well as the list of integers from 1 to 10. The variable k is sequentially assigned each value in the list, which is in turn substituted into the first expression. So the above seq command is equivalent to the sequence of expressions ithprime(1),ithprime(2),...,ithprime(10).

Example 4.3. Generate the elements $p_{100}, \ldots, p_{110}$ of the sequence (4.4).

> seq(isprime(k),k=100..110);

false, true, false, true, false, false, false, true, false, true, false

Thus, 101, 103, 107, and 109 are prime.

Example 4.4. Generate the first 10 elements of the sequence

$$f : k \mapsto \frac{k+1}{k^2} \qquad k \geq 1.$$

We construct a user-defined function for f.

> f:=k->(k+1)/k^2:

> seq(f(n), n=1..10);

$$2,\ \frac{3}{4},\ \frac{4}{9},\ \frac{5}{16},\ \frac{6}{25},\ \frac{7}{36},\ \frac{8}{49},\ \frac{9}{64},\ \frac{10}{81},\ \frac{11}{100}$$

Example 4.5. Generate the set S of the first 30 non-negative powers of 2.

> S:={seq(2^i, i=0..29)};

$$S := \{1, 2, 4, 8, 16, 32, 64, 128, 134217728, 268435456, 536870912,$$
$$256, 512, 1024, 2048, 4096, 8192, 16384, 32768, 65536,$$
$$131072, 262144, 524288, 1048576, 2097152, 4194304,$$
$$8388608, 16777216, 33554432, 67108864\}$$

While the expression sequence produced by seq was in the appropriate order, the enclosure between curly brackets has resulted in a seemingly arbitrary arrangement of the elements of the set.

The `seq` command has the syntax `seq(e,i=a..b)`, where `e` is any expression, `i` is a name, and `a` and `b` are expressions which must evaluate to an integer. The expression `a..b` is of the *range* type

```
> whattype(a..b);
```

$$..$$

Before executing `seq`, Maple expands the range expression, which evaluates to a sequence of integers. This can be done explicitly with the *sequence operator* `$`

```
> $1..10;
```

$$1, 2, 3, 4, 5, 6, 7, 8, 9, 10$$

(The range `a..b` may also be replaced by more general expressions — see chapter 6.)

The value of `seq` is a list of expressions separated by commas, called an *expression sequence*, or, more briefly, a *sequence*. The following command generates a sequence of sets

```
> seq({$1..n},n=0..4);
```

$$\{\}, \{1\}, \{1,2\}, \{1,2,3\}, \{1,2,3,4\}$$

```
> whattype(%);
```

$$exprseq$$

Unlike the elements of a set, the elements of an expression sequence are arranged in a prescribed order. If in the range `a..b` the value of `a` is greater than that of `b`, the result is the NULL sequence. NULL sequences may be defined explicitly, which is useful in initializations

```
> s:=NULL;
```

$$s :=$$

```
> s:=s,1,2,3;
```

$$s := 1, 2, 3$$

Example 4.6. We construct a function `divs(n)` whose value is the set of divisors of a positive integer n (see also `?divisors`). A straightforward (if inefficient) approach consists in applying the function $x \mapsto \gcd(x,n)$ to the set of the first n positive integers. The latter is constructed by expanding the range `1..n` with the operator `$`. The mapping is performed with the function `map`, whose first argument is `igcd`. Because `igcd` is a function of two variables, the second variable must be supplied as the third argument of `map`.

```
> divs:=n->map(igcd,{$1..n},n):
> divs(84);
```

$$\{1, 2, 3, 4, 6, 7, 12, 14, 21, 28, 42, 84\}$$

Evaluating this function requires n calls to the function igcd. We now consider improvements in the efficiency of computation. Apart from the trivial divisors 1 and n, all other divisors of n lie between 2 and $n/2$, so we can save half of the evaluations of igcd

```
> divs:=n->map(igcd,{$2..iquo(n,2)},n) union {1,n}:
```

A more substantial improvement is achieved by computing the divisors d between 2 and $\sqrt{n}$, and then pairing each of them with its twin divisor n/d. To compute $\sqrt{n}$, we use the integer square root function isqrt.

```
> twins:=(x,n)->(igcd(x,n),n/igcd(x,n)):
> divs:=n->map(twins,{$2..isqrt(n)},n) union {1,n}:
```

The number of calls to igcd is now approximately equal to $2\sqrt{n}$. This means that in the case $n = 40000$, the last version of divs is 100 times faster than the first one! However, the function twins still makes two identical calls to igcd; to fix this problem, a more general form of procedures is required (see chapter 9).

Exercises

Exercise 4.1. Consider the sequence $n \mapsto f_n = n/(n+1)$.
(a) Construct a function f(n) whose value is f_n.
(b) Plot the elements $f_0, \ldots, f_{30}$, connecting points with segments.
(c) Using the function f, generate the first 10 elements of the following sequences

(i) $\quad 0, \dfrac{2}{3}, \dfrac{4}{5}, \dfrac{6}{7}, \dfrac{8}{9}, \; \cdots$

(ii) $\quad \dfrac{2}{3}, \dfrac{9}{10}, \dfrac{16}{17}, \dfrac{23}{24}, \; \cdots$

(iii) $\quad \dfrac{1}{2^2}, \dfrac{3^2}{4^2}, \dfrac{5^2}{6^2}, \dfrac{7^2}{8^2}, \; \cdots$

(iv) $\quad 0, \dfrac{3}{4}, \dfrac{8}{9}, \dfrac{24}{25}, \dfrac{35}{36}, \; \cdots$

Exercise 4.2.
(a) Let n be a positive integer. Construct a function S(n) whose value is the set of all rationals between 0 and 1 having the nth prime as denominator (0 and 1 do not belong to S(n)).
(b) Let n be a positive integer, and let $S(n)$ be the set of rationals between 0 and 1 having denominator n. Construct a user-defined function for S.
(c) Verify that $S(12) \cap S(21) = S(3)$.
(d) From numerical experiments, conjecture the value of $S(m) \cap S(n)$, for $m, n \geq 1$.

Exercise 4.3. Construct the set of squares less than 5000 which are not cubes, whence determine its cardinality.

Exercise 4.4*. Let m and n be integers with $m \leq n$, and let $[m, n]$ be an interval which includes the endpoints.

(*a*) Construct a function `findp(m,n)` whose value is 1 if there is a prime in $[m, n]$, and 0 otherwise.

(*b*) Construct a function `nprimes(m,n)` whose value is the number of primes in $[m, n]$.

4.3   Plotting the elements of a sequence

Exploring the behaviour of a sequence graphically is standard practice in experimental mathematics. In this section, we introduce the basic tools of Maple's substantial graphics capabilities, and we apply them to the visualization of sequences.

Maple's graphics are based on the all-purpose function `plot`, which can plot — among other things — a discrete set of points on the plane

$$(x_1, y_1), (x_2, y_2), \ldots, (x_n, y_n).$$

A point (x, y) on the Cartesian plane is represented as a *list* of two elements. This is a new composite data type, consisting of a sequence enclosed in *square* brackets: $[x, y]$. The set of points to be plotted must also be organized in the form of a list, to give

$$[[x_1, y_1], [x_2, y_2], \ldots, [x_n, y_n]].$$

The following commands generate a square, with vertices at the points $(-1, -1)$, $(1, 0)$, $(0, 2)$, $(-2, 1)$.

```
> sq:=[[-1,-1],[1,0],[0,2],[-2,1],[-1,-1]]:
> plot(sq);
```

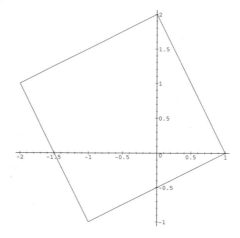

```
> whattype(sq);
```

$$list$$

Maple automatically connects points with a segment, so the vertex $(-1, -1)$ has to be entered twice, or one of the sides would be missing.

Plotting a sequence amounts to plotting points on the plane. Indeed, if $a_1, a_2, \ldots, a_n$ are elements of a sequence, they are represented by the points

$$(1, a_1), (2, a_2), \ldots, (n, a_n)$$

expressed in Cartesian coordinates. For illustration, let us consider the graphical representation of the sequence of gaps between consecutive primes. If p_n is the nth prime we let

$$n \mapsto g_n \qquad g_n = p_{n+1} - p_n, \qquad n = 1, 2, \ldots.$$

We first define a function for the sequence g

```
> g:=n->ithprime(n+1)-ithprime(n):
```

Then we construct the data set corresponding to the first 300 elements of this sequence, assign its value to a variable, and plot it

```
> gap:=[seq([n,g(n)], n=1..300)]:
> plot(gap,style=POINT,title="Gaps between primes");
```

Gaps between primes

The option `style=POINT` plots disconnected points. (Maple's default setting — connecting point with segments — is `style=LINE`). A title has been inserted using the `title` option, which is set equal to an expression enclosed within double quotes (a *string* data type). The gaps are even integers, except for the first one: $g_1 = 3 - 2 = 1$. An n increases, larger gaps appear. Overall, the average gap size seems to increase, but in a very irregular fashion (the study of this phenomenon was a major concern for 19th-century number theorists). The largest gap among the plotted data is located approximately between the 200th and the 250th prime. To locate it precisely, we magnify a portion of the graph, reverting to line style for better visualization.

```
> plot(gap,n=200..250,"g(n)"=15..35);
```

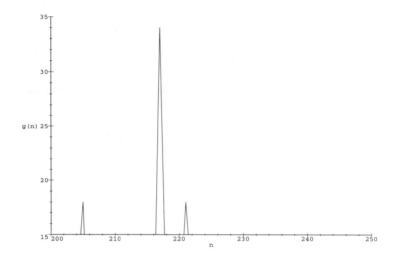

It is clear that the maximum of g occurs for $n = 217$, where the gap size is 34

```
> ithprime(217),ithprime(218);
```

$$1327, 1361$$

The second and third arguments of the call to `plot` define the ranges in the horizontal and vertical axes, respectively, labelling the corresponding axes. The labels must be strings, whence the expression `g(n)` has to be enclosed within double quotes, for it contains parentheses. Any other option (such as a title) must be placed *after* the range specifications (see `?plot` and `?plot[options]` for more information).

4.4 Periodic and eventually periodic sequences

A sequence is *periodic* if it consists of indefinite repetition of the same finite pattern

$$\underbrace{a_0, a_1, \ldots, a_{T-1}},\ \underbrace{a_0, a_1, \ldots, a_{T-1}},\ \underbrace{a_0, a_1, \ldots, a_{T-1}},\ \ldots$$

The number of elements in the repeating block (T in this case) is called the *period* of the sequence. Periodic sequences are of great interest.

A periodic sequence is specified by specifying its repeating block. Let the period be equal to T, and the repeating block $a_0, a_1, \ldots, a_{T-1}$. Periodicity is expressed concisely by the *recursive* notation

$$a_{k+T} = a_k \qquad k \geq 0 \tag{4.5}$$

which says that an arbitrary element of the sequence is equal to the corresponding element in the previous block. Let this element be a_n. By moving backwards from block to block, we eventually reach the first block $a_0, \ldots, a_{T-1}$, so we can identify a_n with one of its elements $a_{n'}$, where n' is an integer lying between 0 and $T - 1$. To compute n' as a function of n, we write $n = qT + r$, where q and r are the quotient and the remainder, respectively, of the division of n by T. Then a_n belongs to the $(q + 1)$th block, and occupies the rth place within the block (starting from $r = 0$). Therefore, $n' = r$, whence $a_n = a_r$.

Example 4.7. Let

$$a : \mathbb{N} \to \{-1, 1\} \qquad a := n \mapsto (-1)^n.$$

This sequence is periodic with period 2:

$$\underbrace{1, -1},\ \underbrace{1, -1},\ \underbrace{1, -1},\ \ldots$$

with the 2-element pattern $1, -1$ repeating indefinitely. This sequence is defined recursively as follows (think about it)

$$a_0 = 1 \quad a_1 = -1 \qquad a_{n+2} = a_n, \qquad n \geq 0.$$

We modify the above sequence into a periodic *binary* sequence, assuming the values 0 and 1, as follows

$$b : \mathbb{N} \to \{0, 1\} \qquad b := n \mapsto \frac{1 + a_n}{2} = \frac{1 + (-1)^n}{2}.$$

Now the repeating block is $1, 0$. Note that this is a representation of the characteristic function of the odd integers.

Example 4.8. Fix a positive integer a. We consider the sequence

$$\gamma : n \mapsto \gcd(n, a) \qquad n \geq 0.$$

We begin with an experiment, with $a = 84$.

```
> plot([seq([n,igcd(n,84)],n=0..200)]);
```

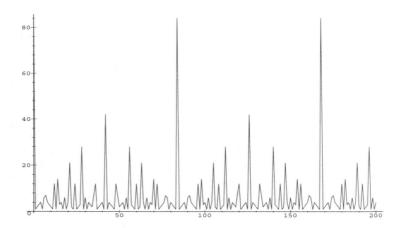

The sequence appears to be periodic with period $a = 84$. Moreover, each repeating block appears to be symmetrical with respect to the midpoint. We now show that this is indeed the case for every value of a. First of all, the sequence γ is periodic with period a, that is, $\gamma_n = \gamma_{n+a}$ for every $n \geq 0$. Indeed, let $\gamma_n = d$. Then $n = dq$ and $a = dq'$, so that $n+a = d(q+q')$. This shows that $n + a$ is also divisible by d. If $n + a$ and a had a common divisor d' greater than d, then d' would also divide their difference $n + a - a = n$, contrary to the hypothesis that d was the greatest common divisor of n and a. Thus, $d = \gcd(n + a, a) = \gamma_{n+a}$. To prove the symmetry property we must establish that $\gcd(n, a) = \gcd(n, n - a)$. The proof of the latter statement is virtually identical to that given above, and will be left as an exercise.

Example 4.9. Let $\gamma : n \mapsto \gcd(n, 5)$. Then γ is periodic with period 5. Thus, $\gamma_{123456789} = \gamma_4 = 1$.

A sequence is *eventually periodic* if it becomes periodic from a certain point onwards:

$$\underbrace{b_0, b_1, \ldots, b_{S-1}}, \underbrace{a_0, a_1, \ldots, a_{T-1}}, \underbrace{a_0, a_1, \ldots, a_{T-1}}, \underbrace{a_0, a_1, \ldots, a_{T-1}}, \ldots$$

An eventually periodic sequence has an initial non-repeating part. The number of its elements is called the *transient length*. The condition (4.5) now becomes

$$a_{k+T} = a_k \qquad k \geq S.$$

Example 4.10. Consider the sequence defined by the characteristic function of the subset $\{0\}$ of $\mathbb{N}$; see (3.16). This sequence is eventually periodic with period 1 and transient length 1.

```
> chi:=n->if n = 0 then 1 else 0 fi:
> seq(chi(i),i=0..5);
```

$$1, 0, 0, 0, 0, 0$$

The quotient and remainder of integer division are functions of two integer variables. By fixing one variable and letting the other run through the integers, we obtain eventually periodic sequences of great interest.

We begin by considering the quotient and the remainder of the division of a fixed integer n by d

$$d \mapsto \mathrm{quo}(n, d) \qquad d \mapsto \mathrm{rem}(n, d), \qquad d = 1, 2, \ldots \quad (n \text{ fixed}).$$

Below, we tabulate the first few elements of these sequences, in the case $n = 5$.

d	1	2	3	4	5	6	7	8	9	10	11	12
$\mathrm{quo}(5, d)$	5	2	1	1	1	0	0	0	0	0	0	0
$\mathrm{rem}(5, d)$	0	1	2	1	0	5	5	5	5	5	5	5

Both sequences become eventually periodic with period 1, after a transient for $1 \leq d \leq 5$. To prove this for an arbitrary value of n, we consider equation (2.10) in the range $d > n$

$$n = d \cdot 0 + n \qquad n < d. \tag{4.6}$$

This shows that in the above range the sequence $d \mapsto \mathrm{quo}(n, d)$ is identically zero. For $d \leq n$ such sequence displays a transient behaviour, decreasing from n (at $d = 1$) to 1 (at $d = n$).

From (4.6) it also follows that the sequence $d \mapsto \mathrm{rem}(n, d)$ is identically equal to n for $d > n$. The transient behaviour in the range $1 \leq d \leq n$ is more complicated, although it still displays a considerable degree of regularity, as illustrated by the figure on the following page, for the case $n = 200$.

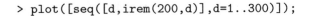

```
> plot([seq([d,irem(200,d)],d=1..300)]);
```

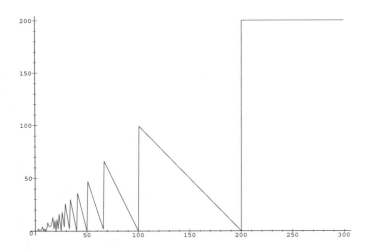

The sequence assumes the value 0 in correspondence to all divisors of n, and its value is small whenever $d = \text{quo}(n, k)$, for $k = 1, 2, \ldots, n$.

4.5   Some non-periodic sequences

Periodic or eventually periodic sequences are, in a sense, atypical. We consider some instances of departure from this behaviour, to gain familiarity with some common patterns.

A simple but important case is that of *linear* rational sequences

$$L : N \to \mathbb{Q} \qquad\qquad n \mapsto a \cdot n \qquad a \in \mathbb{Q}. \qquad (4.7)$$

If $a \neq 0$, then L_n diverges to $+\infty$ or $-\infty$ depending on the sign of a.

Adding a (eventually) periodic sequence to a linear one yields interesting behaviour. These sequences take the form

$$n \mapsto L(n) + P(n) \qquad\qquad (4.8)$$

where P is a (eventually) periodic function. In the special case when P is a constant function, the sequence is said to be *affine.*

For illustration, we consider the quotient and the remainder of the division of n by a fixed integer d

$$n \mapsto \text{quo}(n, d) \qquad n \mapsto \text{rem}(n, d), \qquad n = 0, 1, \ldots, \quad (d \text{ fixed}).$$

Below we tabulate the first few elements of these sequences, in the case $d = 5$.

n	0	1	2	3	4	5	6	7	8	9	10	11	12
quo$(n,5)$	0	0	0	0	0	1	1	1	1	1	2	2	2
rem$(n,5)$	0	1	2	3	4	0	1	2	3	4	0	1	2

The first sequence displays a step-like behaviour, with steps of length 5 and height 1. The second is periodic with period 5; i.e., it consists of indefinite repetition of the same 5-pattern $0, 1, 2, 3, 4$. This behaviour persists when one replaces 5 by an arbitrary value of d. To prove it, we consider the quotient-remainder equation (2.10). Replacing n by $n + d$ we obtain

$$n + d = d(q + 1) + r \qquad 0 \leq r < d, \qquad (4.9)$$

which shows that the quotient increases by 1 every d steps. Because the quotient cannot *decrease* when n increases, then it must increase by 1 precisely once every d steps. Thus, the sequence $n \mapsto$ quo(n, d) has a step-like behaviour, increasing by 1 when n is a multiple of d.

Equations (2.10) and (4.9) show that dividing n and $n + d$ by d gives the same remainder. So the sequence $n \mapsto$ rem(n, d) is periodic, with period at most d. To show that the period is exactly d, we note that as n increases by 1, then r either increases by 1 (if $r \neq d - 1$), or it becomes zero (if $r = d - 1$).

The step-like behaviour of $n \mapsto$ quo(n, d) can be thought of as the result of superimposing a linear and a periodic sequence. From equation (2.10) we obtain

$$q(n) = \frac{1}{d} n - \frac{r(n)}{d}$$

where the n-dependence of q and r has been made explicit. This sequence has the form (4.8), with $a = 1/d$ and $P(n) = -r(n)/d$. The sequence P is periodic with period d, because r is periodic with the same period.

Next we consider the *modulation* of a linear sequence by a periodic one. With reference to (4.7) we have $n \mapsto a(n) \cdot n$, where the function $a = a(n)$ is periodic. As an example, consider the sequence

$$\theta : n \mapsto \text{lcm}(n, d) \qquad (d \neq 0 \text{ fixed}).$$

Because $d \leq \text{lcm}(n, d) \leq nd$, we obtain

$$n \leq \theta_n \leq nd. \qquad (4.10)$$

From (2.28) and the fact that $\gcd(n, d) \neq 0$, we have

$$\text{lcm}(n, d) = n \cdot \frac{d}{\gcd(n, d)}.$$

Because of the periodicity of the greatest common divisor, we see that $n \mapsto \text{lcm}(n, d)$ is the product of a linear and a periodic function.

We wish to plot θ_n, showing graphically the bounds (4.10), for $d = 10$ and $1 \leq n \leq N$. Because the bounding sequences $n \mapsto n$ (lower bound) and $n \mapsto d \cdot n$ (upper bound) are just straight lines, we plot them directly as two solid line segments. The parameters of the problem (the values of d and N) are stored into variables, to gain in generality.

```
> N:=100:
```

```
> d:=10:
```

```
> lower:=[[1,1],[N,N]]:
```

```
> upper:=[[1,d],[N,N*d]]:
```
Next we construct the data set for the sequence θ_n

```
> data:=[seq([n,ilcm(n,d)],n=1..N)]:
```
The three data sets `lower`, `upper`, and `data` must be arranged in a *list* (which is now a list of lists of lists!). The first two are to be represented as solid lines, and the third as a dotted line, as follows

```
> plot([lower,upper,data], linestyle=[1,1,2]);
```

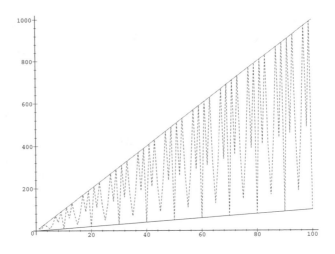

The option `linestyle` is equated to a *list of integers* containing as many elements as the number of data sets to be plotted. The values 1 and 2 correspond to a solid and a dotted line, respectively. The reader is invited to explore the various effects that can be obtained by changing the value of `linestyle`.

Up to now, we have considered non-repeating sequences ranging in an infinite set. But even in a finite set, e.g., $\{0, 1\}$, it is easy to construct ordered patterns that never repeat, such as

$$1, 0, 0, 1, 1, 1, 0, 0, 0, 0, 1, 1, 1, 1, 1, 0, 0, 0, 0, 0, 0, 1, 1, 1, 1, 1, 1, 1, 0, \ldots$$

Equally interesting are non-repeating sequences which display no regularity at all, the *random* sequences. A random sequence seems a contradiction in terms. How can it be random if it can be defined? This is a delicate and interesting problem, which we shall briefly address in the next chapter in connection with random sequences of digits. Here we merely note that finding random-like behaviour in arithmetic is not at all difficult. For instance, a seemingly inexhaustible source of irregular and unpredictable phenomena can be found in the sequence of primes, as the following example illustrates.

Example 4.11.   Let p_n be the nth prime. For $n > 2$, the remainder of the division of p_n by 3 is equal to 1 or 2 (it cannot be equal to zero; otherwise, p_n would be divisible by 3). Consider the following binary sequence

$$\Theta : \ n \mapsto \mathrm{rem}(p_n, 3) - 1 \qquad\qquad n \geq 3.$$

This sequence displays a great deal of irregularity:

```
> theta:=n->irem(ithprime(n),3)-1:
> seq(theta(n),n=3..30);
```

$$1, 0, 1, 0, 1, 0, 1, 1, 0, 0, 1, 0, 1, 1, 1, 0, 0, 1, 0, 0, 1, 1, 0, 1, 0, 1, 0, 1$$

Farey sequences *

For every integer $n \geq 1$, let $\mathcal{F}_n$ be the set of rational numbers in the unit interval, whose denominator does not exceed n

$$\mathcal{F}_n = \left\{ \frac{p}{q} \in \mathbb{Q}, \ \Big| \ 1 \leq q \leq n, \ 0 \leq p \leq q \right\}. \qquad (4.11)$$

Thus,

$$\mathcal{F}_1 = \left\{ \frac{0}{1}, \frac{1}{1} \right\}$$

$$\mathcal{F}_2 = \left\{ \frac{0}{1}, \frac{1}{2}, \frac{1}{1} \right\}$$

$$\mathcal{F}_3 = \left\{ \frac{0}{1}, \frac{1}{3}, \frac{1}{2}, \frac{2}{3}, \frac{1}{1} \right\}$$

$$\mathcal{F}_4 = \left\{ \frac{0}{1}, \frac{1}{4}, \frac{1}{3}, \frac{1}{2}, \frac{2}{3}, \frac{3}{4}, \frac{1}{1} \right\}$$

$$\mathcal{F}_5 = \left\{ \frac{0}{1}, \frac{1}{5}, \frac{1}{4}, \frac{1}{3}, \frac{2}{5}, \frac{1}{2}, \frac{3}{5}, \frac{2}{3}, \frac{3}{4}, \frac{4}{5}, \frac{1}{1} \right\}.$$

We have a nested sequence of finite sets, each properly contained in the following one:

$$\mathcal{F}_1 \subset \mathcal{F}_2 \subset \mathcal{F}_3 \subset \ldots.$$

The set $\mathcal{F}_n$ is called the nth *Farey sequence* (the appellative 'sequence' derives from the fact that the elements of the set $\mathcal{F}_n$ are usually considered in ascending order). For each n and each denominator q in the range $1 \leq q \leq n$, the corresponding numerator ranges from 0 to q, inclusive. In this way, some fractions appear several times, but multiple occurrences are eliminated in accordance with the definition of a set. For instance, $0 = 0/q$ and $1 = q/q$ appear for every value of q, while $1/2$ appears only when q is even.

We wish to construct a user-defined function F(n) whose value is the set $\mathcal{F}_n$. We shall not be concerned with the *ordering* of the elements of $\mathcal{F}_n$ (from the smallest to the largest), and accordingly we shall be using the *set* data type. This will enable us to eliminate repeated entries automatically.

The starting point is to construct an expression sequence containing all fractions with a given denominator q. (We exclude the elements 0 and 1, to limit the number of repeated entries.)

```
> seq(p/q,p=1..q-1):
```

Next we embed the above sequence within an expression sequence which determines the range of values assumed by the denominator q.

```
> seq(seq(p/q,p=1..q-1),q=2..n):
```

Then we add the missing elements 0 and 1, and construct the corresponding set data type.

```
> {0,1, seq(seq(p/q,p=1..q-1),q=2..n)}:
```

Finally, we construct a user-defined function whose value is the above expression.

```
> F:=n->{0,1, seq(seq(k/d,k=1..d-1),d=2..n)}:
> F(5);
```

$$\left\{0, 1, \frac{1}{2}, \frac{1}{3}, \frac{2}{3}, \frac{1}{4}, \frac{3}{4}, \frac{1}{5}, \frac{2}{5}, \frac{3}{5}, \frac{4}{5}\right\}$$

Exercises

Exercise 4.5. Consider the following sequences $n \mapsto a_n$ (p_n denotes the nth prime number)

(1)	$n \mapsto -n^2 + 100$	$n \geq 0$
(2)	$n \mapsto \operatorname{lcm}(n, 6) - 6n$	$n \geq 1$
(3)	$n \mapsto \gcd(n, 2) - \gcd(n, 3)$	$n \geq 1$
(4)	$n \mapsto p_{n+1}(p_n + 1)$	$n \geq 2$
(5)	$n \mapsto$ quotient of division of $3n$ by 7	$n \geq 0$
(6)	$n \mapsto$ remainder of division of p_n by 8	$n \geq 1$

$$(7) \quad n \mapsto \begin{cases} true & \text{if } n^2 + 1 \text{ is prime} \\ false & \text{otherwise} \end{cases} \qquad n \geq 0$$

$$(8) \quad n \mapsto \begin{cases} true & \text{if } n^2 > 7n - 7 \\ false & \text{otherwise} \end{cases} \qquad n \geq 0$$

$$(9) \quad n \mapsto \begin{cases} 0 & \text{if } (a_n + 1)/2 \text{ is even} \\ 1 & \text{otherwise} \end{cases} \quad (a_n \text{ as in part 6}) \quad n \geq 2$$

$$(10) \quad n \mapsto \begin{cases} 1 & \text{if } n \text{ is relatively prime to } 12 \\ 0 & \text{otherwise} \end{cases} \qquad n \geq 1.$$

- For each sequence a_n, construct a user-defined function a(n). Test it for small values of n.

- Display/plot the first few terms, and then display/plot a few terms after the 1000th one.

- From the result of your experiment, formulate a conjecture on the behaviour of the sequence for large n. Try to provide a supporting argument.

Exercise 4.6. The following tongue twister is an exercise in logic. We define
$$M = 10^4 \qquad a_n = n^2 + n + 17, \qquad n \geq 0.$$
(a) Determine the smallest value of n for which a_n is *not* prime.
(b) Determine the smallest value of n for which a_n is prime and greater than M.
(c) Determine the smallest prime value of n for which a_n is greater than M.
(d) Determine the smallest value of n greater than M for which a_n is prime.
(e) Determine the largest prime value of n smaller than $2M$ for which a_n is prime.

Exercise 4.7. Consider the following binary sequence
$$\Lambda_k = \begin{cases} 0 & \text{if } k \text{ is divisible by a square} \\ 1 & \text{otherwise} \end{cases} \qquad k > 1.$$
For instance, $\Lambda(24) = 0$, since $24 = 2^3 \cdot 3$ is divisible by the square 2^2. On the other hand, $\Lambda(26) = 1$, since $26 = 2 \cdot 13$ has no square divisor.
(a) By considering the prime factorization of k, compute Λ_k for $k = 1000, \ldots, 1005$. (Alternatively, you may wish to explore the Maple library numtheory).
(b) Compute the first 5 terms of the sequence
$$\zeta_n = \Lambda_{\theta_n}, \qquad n = 1, 2, \ldots$$
where Λ is as in part (a) and
$$\theta_n = 3^n + 5n, \qquad n = 1, 2, \ldots$$
(Consider the definition of ζ_n carefully.)

Exercise 4.8. Plot the elements of the sequence $d \mapsto \mathrm{rem}(n, d)$ for $n = 100$, and $1 \le d \le 100$. Show that for $d > n/2$, the value of this sequence is $n - d$.

Exercise 4.9. Let p be an odd prime. Use the periodicity of the sequence $n \mapsto \mathrm{rem}(n, 3)$ to show that $p + 2$ and $p + 4$ cannot both be prime.

Exercise 4.10. Let us consider the following sequence

$$ f : \mathbb{N}^+ \to \mathbb{Q} \qquad n \mapsto \frac{8}{8 + n^3} - \frac{1}{3}. $$

(a) By examining the plot of $f(x)$ for x in the range $0 \le x \le 10$, convince yourself that this sequence contains only a finite number of positive elements, and identify them. More precisely, determine an integer k such that

$$ r_k > 0 \qquad\qquad r_{k+t} < 0, \quad t = 1, 2, 3, \ldots \qquad\qquad (4.12) $$

(b) Prove the above statement, with Maple's help. First show that, for the value of k of equation (4.12), $r_k > 0$ and $r_{k+1} < 0$, by direct computation. Then compute the derivative of f using the operator `diff` (find out about it with the online help system), and verify that it is negative for all $x > 0$. Why is this enough? (You may call this a *computer-assisted proof*, albeit a very simple one.)

Exercise 4.11. Let n be a natural number, and let r_n be the remainder of the division of 2^n by 23.

(a) Compute $r_0, \ldots, r_6$ by hand.

(b) Compute r_n with Maple for a sufficient number of small values of n, until you can guess what is the behaviour of the sequence. Describe your findings in the form of a conjecture.

(c) Assuming the validity of your conjecture, compute $r_{123456789}$ (you can't compute it directly, yet!).

Exercise 4.12. We consider the prime decomposition of $n > 1$ given by (2.25). We assume that the primes p_i are all *distinct* and form an *increasing* sequence, that is

$$ p_i < p_{i+1} \qquad\qquad i = 1, 2, \ldots, k - 1. $$

For each n, there are well-defined integers $R_n = k$ (the number of prime divisors of n), $P_n = p_k$ (the largest prime divisor of n), and $E_n = e_k$ (its exponent). (Make sure you understand the meaning of p_k and of e_k, and why the primes in equation (2.25) must be distinct and in ascending order for k, p_k, and e_k to be defined unambiguously. Also note that there is no *explicit* formula expressing R_n, P_n, and E_n as a function of n.)

Compute and tabulate the values of R_n, P_n, and E_n for n equal to the first 5 elements of the sequence

$$ n = n(l) = 5^l - 4^{l-1} \qquad\qquad l = 1, 2, \ldots $$

(Note: Maple does not always display the prime divisors of an integer in ascending order.)

Exercise 4.13. Let r_n be the remainder of the division of the nth prime number by 12 (thus, $r_6 = 1$). It can be shown that for $n \geq 3$, r_n assumes only the four values $1, 5, 7, 11$ (can you prove it?). Write a user-defined function K(n) for the following characteristic function

$$K(n) = \begin{cases} 1 & \text{if } r_n = 1 \text{ or } 11 \\ -1 & \text{if } r_n = 5 \text{ or } 7. \end{cases}$$

Exercise 4.14. Let P_k be the set of primes not greater than k, and let $F_k : P_k \to \mathbb{Z}$ be the function sending x to $x\,(x - k)$. Construct P_k and hence compute $|F_k(P_k)|$, for $k = 2309$ and 2310. For which of these values of k is F_k injective?

4.6 Basic counting sequences

Some important sequences originate from the process of counting elements of certain sets, such as the number of *permutations* of a set of n objects, or the number of *combinations* of n objects, taken k at a time. The process of counting is the main concern of a branch of discrete mathematics called *combinatorics*.

The factorial function and permutations

An important function built from multiplication is the factorial function. If n is a positive integer, the factorial of n, denoted by $n!$, is defined as the product of all positive integers up to and including n, namely

$$n! = 1 \cdot 2 \cdot 3 \cdots (n-1) \cdot n = \prod_{k=1}^{n} k, \qquad n \geq 1.$$

Thus, $5! = 1 \cdot 2 \cdot 3 \cdot 4 \cdot 5 = 120$. From this definition it follows that

$$(n+1)! = 1 \cdot 2 \cdot 3 \cdots n \cdot (n+1) = (1 \cdot 2 \cdot 3 \cdots n) \cdot (n+1) = (n+1)\, n! \qquad n \geq 1.$$

This gives a *recursive* definition of the factorial function — the definition of a function in terms of itself

$$1! = 1; \qquad (n+1)! = (n+1)\, n! \qquad n \geq 1. \tag{4.13}$$

Equation (4.13) is valid only for positive n. We disregard this, and set $n = 0$, which gives

$$1! = 0! \cdot 1 \quad \Longrightarrow \quad 0! = 1.$$

So we have defined the factorial function for all elements of $\mathbb{N}$

$$0! = 1; \qquad (n+1)! = (n+1)\, n! \qquad n \geq 0. \tag{4.14}$$

This result could not have been extracted from the definition (4.13), because for $n = 0$ the product is empty. With this device, we have extended the definition of the factorial function to all non-negative integers. Note that this procedure cannot be used to extend the definition of the factorial to negative integers — try!

Factorials grow very rapidly. Below we compare the factorial function with the exponential functions $2^n, 3^n$, and 4^n, for $n < 10$.

n	0	1	2	3	4	5	6	7	8	9
$n!$	1	1	2	6	24	120	720	5040	40320	362880
2^n	1	2	4	8	16	32	64	128	256	512
3^n	1	3	9	27	81	243	729	2187	6561	19683
4^n	1	4	16	64	256	1024	4096	16384	65536	242144

For any $b > 1$ the function b^n is initially greater than $n!$, but the factorial function eventually takes over. For instance, when $b = 100$ one finds that $n!$ overtakes 100^n at $n = 269$.

What is the combinatorial meaning of the factorial function? We define the concept of a *permutation*. A permutation of n objects in a row is a rearrangement of them. The following is a permutation of $n = 6$ objects, called a, b, c, d, e, f

$$abcdef \quad \mapsto \quad dabfec$$

(Formally, this is just a *bijection* of the set $\{a, b, c, d, e, f\}$.) How many possible permutations of n objects are there? Let us call this number π_n. Clearly $\pi_1 = 1$ and $\pi_2 = 2$. For $n = 3$ we have

$$abc \quad acb \quad bac \quad bca \quad cab \quad cba.$$

Thus, $\pi_3 = 6$. With a little patience, one finds that $\pi_4 = 24$ and the temptation to conjecture $\pi_n = n!$ is irresistible.

This is indeed the case and the proof is simple. There are n ways of choosing the position of the first object. For every choice of the first object there are $n-1$ ways to choose the position of the second, so we have $n(n-1)$ choices of the destination of two objects, and so forth, until we reach the last object for which there is only one choice, because the remaining $n - 1$ objects are already in place.

It is possible to extend the definition of the factorial function to real numbers. The resulting function — called the *gamma function* — will be dealt with in section 5.5.

The binomial coefficient and combinations

Let n and k be positive integers. The binomial coefficient $\binom{n}{k}$ is defined as follows

$$\binom{n}{k} = \frac{n\,(n-1)\,(n-2)\,\cdots\,(n-k+1)}{k!}. \tag{4.15}$$

Thus,

$$\binom{6}{3} = \frac{6 \cdot 5 \cdot 4}{1 \cdot 2 \cdot 3} = 20.$$

One finds immediately from the definition

$$\binom{n}{1} = n \qquad \binom{n}{2} = \frac{n(n-1)}{2}.$$

If $k = 0$ in (4.15), the product at the numerator is empty. In this case, we define

$$\binom{n}{0} = 1.$$

We look at small values of n and k, by displaying *Pascal's triangle*

$$
\begin{array}{c}
1 \\
1, \, 1 \\
1, \, 2, \, 1 \\
1, \, 3, \, 3, \, 1 \\
1, \, 4, \, 6, \, 4, \, 1 \\
1, \, 5, \, 10, \, 10, \, 5, \, 1 \\
1, \, 6, \, 15, \, 20, \, 15, \, 6, \, 1 \\
1, \, 7, \, 21, \, 35, \, 35, \, 21, \, 7, \, 1 \\
1, \, 8, \, 28, \, 56, \, 70, \, 56, \, 28, \, 8, \, 1
\end{array}
\qquad (4.16)
$$

The rows are labelled by the value of n, starting from 0 and proceeding from top to bottom. The elements of the nth row are labelled by k, starting from $k = 0$ and proceeding from left to right up to $k = n$. For example, $\binom{7}{5} = 21$.

We note that the binomial coefficient is an *integer,* which is not at all obvious from definition (4.15). To prove this, we turn to the combinatorial interpretation of the binomial coefficient. A *combination* of n objects taken k at a time is any subset of k objects from a set of n distinct objects (this is where the word combinatorics comes from!). The order in which these objects are arranged makes no difference. For instance, there are 10 combinations of five objects taken two at a time, that is, 10 ways of selecting two objects out of a set of five. If the five objects are taken from the set $\{a, b, c, d, e\}$, then the 10 subsets of 2 elements are

$$\{a,b\}, \, \{a,c\}, \, \{a,d\}, \, \{a,e\}, \, \{b,c\}, \, \{b,d\}, \, \{b,e\}, \, \{c,d\}, \, \{c,e\}, \, \{d,e\}.$$

The number of combinations of n objects taken k at a time is given by the binomial coefficient $\binom{n}{k}$. To see this, we first select k elements and arrange them in a sequence (that is, in a specified order). There are n choices for the first element of the sequence. For each such choice there are $n - 1$ choices for the second element, and so forth, until we choose the kth element, which can be done in $n - k + 1$ different ways. In all, we obtain $n(n-1) \cdots (n-k+1)$ sequences with k elements. But since in a combination the order in which these elements are arranged is irrelevant, we must divide this number by all possible *permutations* of k objects, which we found to be equal to $k!$. So the number of combinations is $n(n-1)\cdots(n-k+1)/k!$, which is the binomial coefficient (see (4.15)).

The binomial coefficient can be represented in terms of factorials as follows

$$\binom{n}{k} = \frac{n(n-1)(n-2)\cdots(n-k+1)}{k!}$$

$$= \frac{n(n-1)(n-2)\cdots(n-k+1)(n-k)!}{k!(n-k)!}$$

$$= \frac{n!}{k!(n-k)!}.$$

From this we obtain

$$\binom{n}{k} = \binom{n}{n-k}. \tag{4.17}$$

If $n < k$, then $n-k+1$ is either 0 or negative. Then, from (4.15)

$$\binom{n}{k} = \frac{n(n-1)\cdots1\cdot0\cdot-1\cdots(n-k+1)}{k!} = 0.$$

Taking into account (4.17), we see that for fixed n, the binomial coefficient can be defined for any integer k, positive or negative, but its value will be nonzero only for $0 \le k \le n$.

The binomial coefficient $\binom{n}{k}$ is implemented in Maple by the function `binomial(n,k)`

```
> evalb(binomial(10,4)=10!/(4!*(10-4)!));
```

$$true$$

To save typing, we can shorten the symbolic name `binomial` to `bc` via the `alias` statement as follows

```
> alias(bc=binomial):
```

This will cause Maple to translate `bc(n,k)` into `binomial(n,k)`.

```
> seq(bc(8,k),k=-2..10);
```

$$0,0,1,8,28,56,70,56,28,8,1,0,0$$

The name binomial coefficient derives from the following result, called the *binomial theorem*

Theorem 2

$$(x+y)^n = \sum_{k=0}^{n} \binom{n}{k} x^k y^{n-k} \qquad n \ge 0.$$

To see the binomial theorem in action, we expand the expression $(x + y)^8$ (cf. (4.16))

```
> expand((x+y)^8);
```

$$x^8 + 8\,x^7y + 28\,x^6y^2 + 56\,x^5y^3 + 70\,x^4y^4 + 56\,x^3y^5 + 28\,x^2y^6 + 8\,xy^7 + y^8$$

Setting $x = y = 1$ and $x = 1, y = -1$ in Theorem (2), respectively, we obtain the well-known identities

$$\sum_{k=0}^{n} \binom{n}{k} = 2^n \qquad \sum_{k=0}^{n} (-1)^k \binom{n}{k} = 0.$$

Exercises

Exercise 4.15.

(a) Determine how many different football teams (11 players) can be made out of 30 players.

(b) Define a function row(n) whose value is the nth row of Pascal's triangle, in the form of an expression sequence. Thus,

```
> row(3);
```

$$1, 3, 3, 1$$

(c) Using row generate the rows from 0 to 7 of Pascal's triangle.

(d) In Pascal's triangle, observe the symmetry induced by the identity

$$\binom{n}{k} = \binom{n}{n-k}$$

which causes every entry to appear exactly twice, except when n is even, in which case there is a unique central element. Construct the sequence of the central elements of the even rows of Pascal's triangle

$$c_m = \binom{2m}{m} \qquad m \geq 0.$$

Compute c_{15}, c_{50}.

(e) Find an explicit formula for c_m, involving factorials. Check your formula with Maple, in the case $m = 200$.

(f) Plot the coefficients of $(1 + x)^{30}$. Convince yourself that the value of the central coefficient as inferred from the plot is consistent with what you have computed in part (d).

(g) From the plot alone, determine how many coefficients are larger than 10^8, identifying explicitly the indices of the corresponding binomial coefficients. Then compute all the coefficients that are greater than 10^8 with Maple.

Exercise 4.16. The following identity holds

$$\sum_{s=k}^{n} \binom{s}{k} = \binom{k}{k} + \binom{k+1}{k} + \cdots + \binom{n-1}{k} + \binom{n}{k} = \binom{n+1}{k+1}. \quad (4.18)$$

(*a*) Verify equation (4.18) by hand, in the cases $k = 0, n = 2$ and $k = 1, n = 2$.
(*b*) Verify equation (4.18) in the cases $k = 9, n = 13$ and $k = 1000$, $n = 1002$.

4.7   Sequences defined recursively

In this chapter, we have dealt mainly with sequences given by *explicit* functions of the integers

$$x_t = an \; explicit \; function \; of \; t \qquad t \geq 1.$$

We have pointed out that this construction is not always possible. However, how can one characterize the structure of a sequence whose terms cannot be defined explicitly? A very interesting special case is that of sequences defined *recursively*, which in the simplest instance have the structure

$$x_t = an \; explicit \; function \; of \; x_{t-1} \qquad t \geq 1, \quad (4.19)$$

that is, each element of the sequence is defined as an explicit function of the previous element. A sequence of this type, called a *first-order recursive sequence*, is uniquely defined once we specify the first element x_0, which is called the *initial condition*.

Specifically, let us consider a set A, an element $\alpha \in A$, and a function $f : A \to A$. The recursive sequence generated by f with initial condition α is defined as follows

$$x_0 = \alpha \qquad x_t = f(x_{t-1}), \qquad t \geq 1.$$

The first few terms of this sequence are

$$x_0 = \alpha, \quad x_1 = f(x_0) = f(\alpha), \quad x_2 = f(x_1) = f(f(\alpha)), \; \ldots$$

In a recursive sequence, the rule that associates to the integer t the element x_t is: "apply f to x_0 t times".

Example 4.12. Let

$$f : \mathbb{Z} \to \mathbb{Z} \qquad f := x \mapsto -2x + 1.$$

The corresponding recursive sequence with initial condition $x_0 = -1$ is given by

$$x_0 = -1, \quad x_1 = f(-1) = 3, \quad x_2 = f(3) = -5, \quad x_3 = f(-5) = 11, \quad \ldots$$

In Maple, there is no need for subscripts
```
> f:=z->-2*z+1:
> x:=-1;x:=f(x);x:=f(x);x:=f(x);
```

$$x := -1$$
$$x := 3$$
$$x := -5$$
$$x := 11$$

We can even dispose of variables altogether
```
> -1;f(%);f(%);f(%);
```

$$-1$$
$$3$$
$$-5$$
$$11$$

It should be clear that the sequence changes by changing initial conditions
```
> -2;f(%);f(%);f(%);
```

$$-2$$
$$5$$
$$-9$$
$$19$$

Example 4.13. Let Ω_2 be the set of all words of the two letters a and b (see (3.5)). Let $w \in \Omega_2$. We define a function $f : \Omega_2 \to \Omega_2$ whose value at w is the word whose letters are obtained from those of w by replacing every a with b and every b with ab. Thus, if $w = aaba$ then $f(w) = bbabb$. Then we have the recursive sequence

$$w_0 = w \qquad w_{t+1} = f(w_t), \qquad t \geq 0.$$

For instance, the initial condition $w_0 = a$ gives

$$a, \ b, \ ab, \ bab, \ abbab, \ bababbab, \ \ldots \qquad (4.20)$$

To implement the above construct in Maple, we represent a word as a *list* of the symbols a and b
```
> w:=[a,b,b,b]:
```
Constructing the function f requires *simultaneous* substitutions, obtained by enclosing the substitution list between curly brackets (see section 2.3)
```
> f:=w->subs({a=b,b=(a,b)}, w):
```
(Persuade yourself that sequential substitutions would not work in this case.) Note that the second argument in the substitution list required the use of round brackets: without them the substitution list would be interpreted as a=b,b=a,b, resulting in an error termination when the function is called. Thus,

```
> f([a,b,b]);
```

$$[b, a, b, a, b]$$

The construction of the recursive sequence (4.20) now proceeds as usual, e.g.,

```
> [a]:f(%):f(%):f(%):f(%):f(%);
```

$$[b, a, b, a, b, b, a, b]$$

In the general case, a sequence defined recursively has the following structure

$$x_t = f(x_{t-1}, x_{t-2}, \ldots, x_{t-s}, t) \qquad\qquad t \geq s. \qquad (4.21)$$

Here f is an arbitrary function of $x_{t-1}, x_{t-2}, \ldots, x_{t-s}$ as well as of t. Such a sequence is specified by assigning the s initial values $x_0, x_1, \ldots, x_{s-1}$. The integer s is called the *order* of the recursion. If the function f in (4.21) depends on t, the recursive sequence is said to be *non-autonomous*.

Example 4.14. The factorial sequence (4.14) is a first-order non-autonomous recursive sequence. With the notation of (4.21), we have $f(x_{t-1}, t) = t \cdot x_{t-1}$.

Example 4.15. *The Fibonacci numbers.* We consider the recursive sequence

$$F_0 = 1 \qquad F_1 = 1 \qquad F_{t+1} = F_t + F_{t-1}, \quad t \geq 1.$$

With reference to (4.21), we have $f(x_{t-1}, x_{t-2}) = x_{t-1} + x_{t-2}$, and therefore this is a second-order autonomous recursive sequence ($s = 2$ and no dependence of f on t). The first few elements of this sequence are displayed below

t	0	1	2	3	4	5	6	7	8	9	10	11	12	$\cdots$
F_t	0	1	1	2	3	5	8	13	21	34	55	89	144	$\cdots$

The numbers F_t are known as *Fibonacci numbers* (named after the 13th-century mathematician Leonardo da Pisa, nicknamed Fibonacci). They have several important combinatorial applications. We compute F_{10} with Maple

```
> 0:1:%+%%:%+%%:%+%%:%+%%:%+%%:%+%%:%+%%:%+%%:%+%%;
```

$$55$$

Maple has a library function generating the Fibonacci numbers. For details, see ?fibonacci.

Example 4.16. We prove by induction the following identity (discovered by Cassini in 1680)

$$F_{n+1}F_{n-1} - F_n^2 = (-1)^n \qquad\qquad n \geq 1. \qquad (4.22)$$

Equation (4.22) is true for $n = 1$. Assume it true for an arbitrary $n \geq 1$. Then

$$\begin{aligned} F_{n+2}F_n - F_{n+1}^2 &= (F_{n+1} + F_n)F_n - F_{n+1}(F_n + F_n - 1) \\ &= F_{n+1}F_n + F_n^2 - F_{n+1}F_n - F_{n+1}F_{n-1} \\ &= -(-1)^n = (-1)^{n+1}. \end{aligned}$$

This shows that (4.22) is valid for all positive integers n.

Exercises

Exercise 4.17.   The factorial sequence $n!$ was defined recursively in (4.14). We define the analogous first-order non-autonomous recursive sequence $n?$ as follows

$$0? = 1 \qquad (n + 1)? = \left(n + \frac{1}{2}\right) \cdot n? \qquad n \geq 0.$$

Compute $6?$.

Exercise 4.18.   Consider the following second-order autonomous sequence — cf. (4.21)

$$f_0 = 0 \qquad f_1 = 1 \qquad f_{n+1} = 11 \cdot f_n + f_{n-1}, \qquad n \geq 1.$$

(Note that this sequence is determined by *two* initial conditions, f_0 and f_1.) Show that f_6 and f_7 are relatively prime.

Exercise 4.19.   Let r be a rational number, and let f be given by

$$f : \mathbb{Q} \to \mathbb{Q} \qquad f := x \mapsto -\frac{3}{2}x^2 + \frac{5}{2}x + 1.$$

We consider the first-order recursive sequence of rational numbers

$$r_0 = r \qquad r_{t+1} = f(r_t), \qquad t \geq 0. \tag{4.23}$$

(a)   Show that if r_t is an integer, so is r_{t+1}. [*Hint:* consider the cases r_t even and odd separately.]

(b)   Let $r_0 = 4$. Show that r_4 is divisible by 857, hence determine the smallest value of t for which $r_t < -10^{40}$.

(c)   Let $r_0 = 4/3$. Show that for $t \geq 2$ the sequence repeats indefinitely, and compute its period (the length of the repeating part). Hence compute r_{1000}.

Exercise 4.20.   Consider the recursive sequence (4.23), with f given by

$$f : \mathbb{Q} \to \mathbb{Q} \qquad f := x \mapsto \frac{3x + 1}{x + 3}.$$

(*a*) Construct the Boolean characteristic function χ of the set of rationals whose distance from 1 is less than $1/100$.

(*b*) Consider the sequence corresponding to the initial condition $r_0 = 1/2$. Convince yourself that r_t approaches the point $x = 1$. Hence, using the function χ, determine the smallest value of t for which the distance between r_t and 1 is smaller than $1/100$.

(*c*) Prove that if $r_0 = 1$, then $r_t = 1$ for all positive t.

(*d*) Construct the function `fp(r)` whose value is the fractional part of r.

(*e*) Let $\mathbb{Q}_I$ denote the set of rational numbers lying between 0 and 1. Thus, $\mathbb{Q}_I$ consists of all rationals $r = m/n$ with $0 < m < n$. Consider the function

$$g : \mathbb{Q}_I \to \mathbb{Q}_I \cup \{0\} \qquad g := r \mapsto \left\{ \frac{1}{r} \right\}. \qquad (4.24)$$

Prove that g is indeed a function of $\mathbb{Q}_I$ into $\mathbb{Q}_I \cup \{0\}$; that is, prove that if $r \in \mathbb{Q}_I$, then $g(r) \in \mathbb{Q}_I \cup \{0\}$.

(*f*) Using the function `fp` developed above, construct the function `g(r)` given by (4.24).

(*g*) Consider the following recursive sequence

$$r_0 = r \qquad r_{t+1} = \begin{cases} g(r_t) & r_t \neq 0 \\ 0 & \text{otherwise} \end{cases} \qquad t \geq 0. \qquad (4.25)$$

Let $r = 11/19$. Determine the smallest value of t for which r_t is equal to zero.

(*h*) Choose an arbitrary rational number $r \in \mathbb{Q}_I$. Verify that the corresponding sequence r_t eventually reaches zero.

(*i*) Prove that for any choice of $r \in \mathbb{Q}_I$, the sequence r_t with initial condition $r_0 = r$ reaches 0 in a finite number of steps.

Exercise 4.21*. Let the sequence r_t be as in (4.25). From the previous problem it follows that to every rational r in Q_I we can associate the natural number $\tau(r)$, which is the number of iterations needed to reach zero starting from r

$$\tau : \mathbb{Q}_I \to \mathbb{N} \qquad \tau := r \mapsto \text{smallest } t \text{ for which } r_t = 0, \text{ and } r_0 = r.$$

For instance, $\tau(4/7) = 3$, because

$$r_0 = \frac{4}{7} \ \to \ r_1 = g(r_0) = \frac{3}{4} \ \to \ r_2 = g(r_1) = \frac{1}{3} \ \to \ r_3 = g(r_2) = 0.$$
$$(4.26)$$

Now let $\mathcal{F}_n$ be the nth Farey sequence, defined in (4.11). Then $\mathcal{F}_n$ is a finite subset of $\mathbb{Q}_I$. Let Γ_n be the maximum value attained by $\tau(r)$ as r scans all the elements of the set $\mathcal{F}_n$. For example, since $3/4 \in \mathcal{F}_5$, from the data (4.26) we obtain the estimate $\Gamma_5 \geq 2$ (in fact, $\Gamma_5 = 3$).

Find a good estimate for Γ_n, in the case $n = 10000$. Specifically, obtain an estimate $\Gamma_{10000} \geq \gamma$ by producing a rational r in $\mathcal{F}_{10000}$ for which $\gamma = \tau(r)$ is as large as possible. You must display the calculations showing that the sequence with initial condition r reaches zero in γ steps. [*Hint:* The set $\mathcal{F}_{10000}$ has 30397486 elements. Careful experimentation and thinking will get you much further than blind trial and error.]

Exercise 4.22. In this problem, we compare the growth rate of the exponential sequence 2^n with that of the power sequence n^b, where b is a given integer greater than 1, and $n = 1, 2, \dots$. Both sequences diverge to infinity, but the exponential 'gets there first'; that is, it grows faster than any power. However, the sequence of powers starts off faster than the exponential (for $n > 1$) but is eventually overtaken by the latter. This phenomenon — illustrated below for the cases $b = 2$ and 3 — becomes more pronounced as b gets large.

n	1	2	3	4	5	6	7	8	9	10
2^n	2	4	8	16	32	64	128	256	512	1024
n^2	1	4	9	16	25	36	49	64	81	100
n^3	1	8	27	64	125	216	343	512	729	1000

Ignoring the value $n = 1$, we see that 2^n overtakes n^2 at $n = 5$, and n^3 at $n = 10$.

For fixed $b > 1$, let O_b be the smallest $n > 1$ for which $2^n > n^b$

$$O_b : n \mapsto \text{smallest } n \text{ for which } 2^n > n^b \qquad n > 1, \quad b \geq 1.$$

Thus, O_b is the integer at which 2^n overtakes n^b, whence $O_2 = 5$ and $O_3 = 10$, from the above data. Compute and tabulate O_b for $b = 2, \dots, 10$. Display only the calculations that suffice to establish your result.

Exercise 4.23. In this problem, we compare the behaviour of the factorial function $n!$ with that of the exponential function b^n, where b is a given positive integer. Convince yourself that for any fixed $b \geq 1$, $n!$ is smaller than (or equal to) b^n for sufficiently small n. However, the factorial function is known to grow faster than any exponential, so at some point $n!$ will overtake b^n.

Let O_b be the point at which the factorial overtakes the powers of b. So we have an integer sequence $O_1, O_2, O_3, \dots$. Compute and tabulate O_b for $b = 2, \dots, 10$. Display only the calculations that suffice to establish your result.

Exercise 4.24. Consider the following recursive sequence

$$\alpha_0 = 1 \qquad \alpha_{k+1} = f(\alpha_k) = \alpha_k^2 + 1, \qquad k \geq 0.$$

(*a*) Show that α_6 is divisible by 1277.

(*b*) Show that α_9 has 91 decimal digits, without using `length`. [*Hint:* how do you show that 88 has two decimal digits?]

(*c*) This is the fastest-growing sequence you have come across so far. It grows faster than the factorial sequence. Indeed, α_k eventually overtakes $(k!)^b$ for any positive integer b. Let O_b be the point at which α_k overtakes $(k!)^b$. Compute O_1 and O_2.

Exercise 4.25. Construct the fastest-growing sequence you can think of. Specifically, replace the 10 question marks in the following function definition

```
> fast:=n->??????????;
```

with *at most* 10 characters so as to obtain a Maple function `fast(n)`, representing a rapidly growing sequence defined for all positive n. (Any character that is on the keyboard is allowed, but you cannot make use of functions you have defined elsewhere.)

What counts is the asymptotic behaviour of your sequence (how fast it grows when n becomes very large). For instance, $n \to n^2$ starts slower than $n \to n + 10^{10}$, but it gets to infinity first.

Chapter 5

Real and complex numbers

Every rational number is representable as a *pair* of integers, the numerator and the denominator. Are all numbers rational? If not, can at least all numbers be represented as a *finite* collection of integers? The answer to both questions turns out to be no, but a satisfactory explanation is far from simple. There are a lot more numbers than the rationals, and the numbers that are representable as a finite collection of integers are still extremely rare (but also extremely interesting). Yet, based on the limited evidence we have gathered so far, it is hard to be convinced that almost all numbers are *not* rational. For instance, if we represent the rationals as points on a line, we can find them as close as we please to any given point.

This chapter is devoted to answering our first question. To do so we must first introduce a new way of representing numbers, the representation in terms of decimal digits.

5.1 Digits of rationals

One of the best-known numbers is π

$$\pi = 3.141592\ldots$$

This notation is shorthand for a sum

$$
\begin{aligned}
\pi \ =\ & 3 \\
+\ & 0.1 \\
+\ & 0.04 \\
+\ & 0.001 \\
+\ & 0.0005 \\
+\ & 0.00009 \\
+\ & 0.000002 \\
& \ \ \vdots
\end{aligned}
$$

Each term in the sum is *rational*

$$
\pi = 3 \cdot 10^0 + \frac{1}{10^1} + \frac{4}{10^2} + \frac{1}{10^3} + \frac{5}{10^4} + \frac{9}{10^5} + \frac{2}{10^6} + \cdots \qquad (5.1)
$$

and one sees that the decimal point (or *radix point*) separates the non-negative powers of 10 from the negative ones. Can we conclude that π — as a sum of rationals — is itself rational? Not necessarily.

We know that $\mathbb{Q}$ is closed with respect to addition; i.e., the sum of two rationals is still rational. From this property we can conclude that the sum of three rationals is rational, and so on. Therefore, the sum of any *finite* number of rationals is rational, but the sum (5.1) contains an *infinite* number of terms! But equally, this fact alone is not sufficient to demonstrate that a number is *not* rational. For instance, the decimal expansion of 1/3

$$
\frac{1}{3} = 0.33333\ldots = \frac{3}{10^1} + \frac{3}{10^2} + \frac{3}{10^3} + \cdots
$$

is structurally identical to that of π, but in this case the value of an infinite sum of rationals is rational.

Before clarifying this point we develop the Maple tools for generating decimal expansions. Maple provides the standard library function `evalf`, which turns a number into a string of decimal digits. The default number of digits is equal to ten, which is the value of a Maple constant called `Digits`.

```
> Digits;
```
$$
10
$$

```
> Pi,evalf(Pi);
```
$$
\pi,\ 3.141592654
$$

```
> Digits:=20:
> evalf(Pi),evalf(Pi,3);
```
$$
3.1415926535897932385,\ 3.14
$$

Note the possibility of overriding temporarily the default value of `Digits`, with the optional second argument of `evalf`.

We remark that for a given accuracy, not all displayed digits are necessarily exact, since they are chosen in such a way as to minimize the error. For instance, the tenth decimal digit of π is 3, but since π is closer to 3.141592654 than to 3.141592653, when `Digits` is set equal to 10, the last displayed digit is 4. The following example is even more dramatic

```
> 0.099991:
> %,evalf(%,4),evalf(%,3);
```

$$.099991, .09999, .100$$

When the accuracy is reduced to three digits, *none* of the displayed digits is correct!

We begin with some experiments on digits of rationals. As we need to gather clear evidence, we set a large value of the constant `Digits`

```
> Digits:=50:
> 9/176:%=evalf(%);
```

$$\frac{9}{176} = .051136363636363636363636363636363636363636363636364 \quad (5.2)$$

```
> 11/101:%=evalf(%);
```

$$\frac{11}{101} = .10891089108910891089108910891089108910891089108911 \quad (5.3)$$

```
> 999999/1048576:%=evalf(%);
```

$$\frac{999999}{1048576} = .95367336273193359375000000000000000000000000000000 \quad (5.4)$$

```
> 15/17:%=evalf(%);
```

$$\frac{15}{17} = .88235294117647058823529411764705882352941176470588 \quad (5.5)$$

The common feature of these decimal expansions is the existence of a block of digits that appears to repeat indefinitely. This block is 36 in (5.2), 1089 in (5.3), 0 in (5.4), and 8823529411764705 in (5.5). Repetition does not necessarily set in from the very beginning, but it may be preceded by a non-repeating block, which is 0511 in (5.2) and 95367336273193359375 in (5.4). In the remaining examples, there is no such non-repeating block. These observations can be summarized by saying that the digits of a rational number appear to be *eventually periodic* (see section 4.4). This is indeed true, as detailed in the next two theorems.

Theorem 3 *A number is rational if and only if its decimal digits are eventually periodic.*

We analyze the statement of this theorem in some detail. We know that the elements of $\mathbb{Q}$ are ratios of integers. Let us denote by P the set of all numbers whose decimal expansion is eventually periodic. The theorem says that $P = \mathbb{Q}$, thereby establishing an alternative way of defining the set $\mathbb{Q}$.

This statement consists of a two-way implication, expressed with the condition 'if and only if'. The first implication, 'a number is rational *if* its digits are eventually periodic' means that eventual periodicity implies rationality

$$x \text{ is rational} \quad \Longleftarrow \quad \text{the digits of } x \text{ are eventually periodic}$$

which we write formally as

$$x \in \mathbb{Q} \quad \Longleftarrow \quad x \in P.$$

By definition, this means that $P \subseteq \mathbb{Q}$ (cf. (3.7)). Note that stating that P is a subset of $\mathbb{Q}$ does not necessarily imply that P is *properly contained in* $\mathbb{Q}$. It just means that every element of P is also an element of $\mathbb{Q}$.

The *if* implication alone does not rule out the possibility that some rational number has non-repeating digits. To ensure that such a possibility does not actually occur, we need the converse implication 'a number is rational *only if* its decimals are eventually periodic', which states that rationality implies eventual periodicity

$$x \text{ is rational} \quad \Longrightarrow \quad \text{the digits of } x \text{ are eventually periodic}$$

or, equivalently,

$$x \in \mathbb{Q} \quad \Longrightarrow \quad x \in P,$$

which now says that $\mathbb{Q} \subseteq P$ (cf. (3.8)). The two inclusion relations put together give

$$x \in \mathbb{Q} \quad \Longleftrightarrow \quad x \in P$$

which is the definition of equality between sets: $P = \mathbb{Q}$. We shall give a constructive proof of the $\Longleftarrow$ implication below. The proof of $\Longrightarrow$ will be proposed as an exercise.

The next result provides detailed information on the structure of the digits of a rational.

Theorem 4 *Let $r = p/q$ be a rational number with p and q coprime and $0 \leq r \leq 1$. Then the decimal expansion of r consists of m pre-periodic digits followed by n digits which repeat themselves indefinitely. Letting*

$$q = 2^a 5^b R \qquad \gcd(10, R) = 1$$

we have that $m = \max(a, b)$. As to n, we have two cases:

(i) If $R > 1$, then n is the smallest integer for which 10^n gives remainder 1 when divided by R. (In particular, $n < R$).

(ii) If $R = 1$, then $n = 1$, and r has two decimal representations, where the repeated digit is 0 and 9, respectively.

Note that case (ii) corresponds to the denominator q being divisible only by 2 or 5. We can see the theorem in action in the examples we have illustrated above. Take the case $p/q = 9/176$. Numerator and denominator are coprime, or Maple would have eliminated any common factor. We factor the denominator

> ifactor(176);

$$(2)^4 \, (11)$$

and in the notation of Theorem 4, we have $a = 4, b = 0$, and $R = 11$. Thus, $m = 4$, and $n < 11$. To determine the actual value of n, we consider powers of 10, and note that $10^1 = 0 \cdot 11 + 10$ and $10^2 = 9 \cdot 11 + 1$, so 2 is the smallest power of ten which gives remainder 1, modulo 11, whence $n = 2$.

Computing the digits of a rational

We describe an algorithm for computing the digits of a rational number $r = p/q$, which makes use only of the arithmetic in $\mathbb{Q}$. Without loss of generality we may assume that $0 \le r < 1$. We begin by writing out the decimal expansion of r

$$r \; = \; \frac{p}{q} \; = \; 0.d_1 d_2 d_3 \cdots \qquad\qquad 0 \le d_k < 10, \quad k = 1, 2, \cdots. \qquad (5.6)$$

We are interested in identifying the nth digit d_n of r. Multiplying r by 10^n amounts to shifting the decimal digits of r to the left by n places

$$10^n r \; = \; d_1 d_2 \cdots d_n . d_{n+1} d_{n+2} d_{n+3} \cdots \; = \; d_1 d_2 \cdots d_n + 0.d_{n+1} d_{n+2} d_{n+3} \cdots$$

The integer $d_1 d_2 \cdots d_n$ is the integer part of $10^n p/q$, namely the quotient of the integer division of $10^n p$ by q. Dividing such an integer part by 10 leaves remainder d_n, which is the desired digit (think about it). Thus,

> dgt:=(r,n)->irem(iquo(10^n*numer(r),denom(r)),10):
> evalf(5/13);(seq(dgt(5/13,n),n=1..10));

$$.3846153846$$
$$3, 8, 4, 6, 1, 5, 3, 8, 4, 6$$

A variant of the above algorithm provides a constructive proof that the digits of a rational must eventually repeat. With reference to equation (5.6), we multiply r by 10, causing the first digit d_1 to be moved to the left of the radix point

$$10r \; = \; d_1 . d_2 d_3 d_4 \ldots$$

Then d_1 — the first digit of r — is the integer part of $10r$. To extract the second digit of r we consider the number

$$r_2 \; = \; 10 \cdot r - d_1 \; = \; 0.d_2 d_3 d_4 \ldots$$

and then repeat the same process as before.

We carry out this iterative procedure explicitly, and show that it eventually leads to repetition. Let $r = r_1 = p_1/q$. Then the condition $0 \le r < 1$ implies $0 \le p_1 < q$, and

$$10r_1 \;=\; \frac{10p_1}{q} \;=\; d_1 + \frac{p_2}{q} \;=\; d_1 + r_2.$$

The integer p_2 — the remainder of division of $10r$ by q — is uniquely determined by p_1, and $0 \le p_2 < q$. Repeating this process, we obtain a sequence of integers $p_1, p_2, \ldots$, all lying between 0 and q, and such that p_k is uniquely determined by its predecessor p_{k-1}.

In other words, the sequence $p_1, p_2, \ldots$ is a *recursive sequence* of elements of the set $D = \{0, 1, 2, \ldots, q - 1\}$, whose initial condition p_1 is the numerator of the rational in question. But since D has a *finite* number of elements, after at most $|D| = q$ iterations the integer p_k will necessarily assume a value assumed before (why?). From that point on, its values will begin to repeat, and with them the digits of r.

As an example, we compute the digits of 1/7.

$$
\begin{aligned}
\frac{10 \cdot 1}{7} &= 1 + \tfrac{3}{7} \\
\frac{10 \cdot 3}{7} &= 4 + \tfrac{2}{7} \\
\frac{10 \cdot 2}{7} &= 2 + \tfrac{6}{7} \\
\frac{10 \cdot 6}{7} &= 8 + \tfrac{4}{7} \\
\frac{10 \cdot 4}{7} &= 5 + \tfrac{5}{7} \\
\frac{10 \cdot 5}{7} &= 7 + \tfrac{1}{7}
\end{aligned}
$$

Thus, the repeating digits of 1/7 are 142857, without a non-periodic head. We implement the above recursion in Maple

```
> p:=1:q:=7:iquo(10*p,q);
```
$$1$$

```
> p:=irem(10*p,q):iquo(10*p,q);
```
$$4$$

```
> p:=irem(10*p,q):iquo(10*p,q);
```
$$2$$

The above computation is inefficient, because in a single call to `iquo` or `irem`, Maple computes *both* quotient and remainder. A nice way of solving this problem is to call `iquo` with the optional *third* argument 'p' (see section 2.7). In this way the value of `irem(10*p,q)` is computed and automatically assigned to p.

```
> p:=1:q=7:
> seq(iquo(p*10,q,'p'),k=1..7);
```

$$1, 4, 2, 8, 5, 7$$

Exercises

Exercise 5.1. Consider the following rational numbers r

$$(i) \quad \frac{10001}{11110} \qquad (ii) \quad \frac{13}{23} \qquad (iii) \quad \frac{1}{585728}.$$

Let m be the number of the pre-periodic decimals, and n the period of the repeating decimals of r.

(a) For each value of r, evaluate a sufficient number of decimal digits, and hence conjecture the values of m and n. Identify the pre-periodic head of m digits, as well as the n repeating digits of the tail.

(b) Express the denominator of r in the form $2^a 5^b R$, where R is relatively prime to 2 and 5. Compute m from a and b, and check that for $R > 1$ you have $n < R$.

(c) Summarize your results in a table.

(d) For the case (ii) determine the third digit of r using only the functions `irem` and `iquo`.

Exercise 5.2. Let $x = a/b$ and $y = c/d$ be rational numbers, in reduced form. Their *midpoint* and *mediant*, given by $(x + y)/2$ and $(a + c)/(b + d)$, respectively, lie in $[x, y]$ — the interval with endpoints x and y (cf. equations (2.20) and (2.21)).

(a) Construct a function `mid(x,y)` whose value is the midpoint of x and y.

(b) Construct a function `med(x,y)` whose value is the mediant of (the reduced form of) x and y.

(c) Let $\beta = \sqrt{2} - 1$. (You may assume that β is not rational.) Show that β lies in $[0, 1]$.

(d) Compute the midpoint r of 0 and 1, using `mid`, hence decide which of the two sub-intervals $[0, r]$ or $[r, 1]$ contains β. (Convince yourself that $\beta \neq r$, because r is rational and β is not.) Then compute the midpoint of that interval, and decide which of the two resulting sub-intervals contains β, and so on. Repeat this process until you have determined a rational number whose distance from β is less than 10^{-2}.

(e) Do the same with the mediant in place of the midpoint, using `med`. Compare the performance of the two methods with a short comment.

Exercise 5.3*. Prove that if the digits of a number α are eventually periodic, then one can find integers k and m such that $\alpha \, 10^k + m$ has periodic digits.

Exercise 5.4*.   Prove that a number whose digits are eventually periodic is rational. [*Hint:* first prove it in the periodic case, then use the result of the previous exercise.]

5.2   Real numbers

In the previous section we have identified the distinctive feature of the digits of a rational number, their eventual periodicity. This gives us a constructive method for building *irrational numbers*. All we have to do is to produce strings of digits that do not repeat, some examples of which are given below (these examples, however, turn out to be atypical, as we shall see)

$$\alpha_1 = 0.1234567891011121314151617181920212324252627282930\ldots$$
$$\alpha_2 = 0.40400400040000400000400000040000000400000000040000\ldots$$
$$\alpha_3 = 0.12233344445555566666677777778888888899999999991010\ldots$$

$$(5.7)$$

Putting together the rational numbers (eventually periodic digits) and the irrational numbers (non-eventually periodic digits), we obtain the set $\mathbb{R}$ of *real numbers.* Such set is the collection of all decimal expansions having a finite number of digits to the left of the radix point. Any such expansion defines an infinite sum whose value is the real number in question (see exercises). It should be remembered that the correspondence between digit strings and real numbers fails to be bi-unique in the case of rationals whose denominator is divisible only by 2 or 5, which have two distinct expansions. This ambiguity gives rise to quite a lot of problems in *numerical analysis,* the branch of computing concerned with manipulating real numbers.

Because $\mathbb{R}$ contains $\mathbb{Q}$, we have the chain of inclusions

$$\mathbb{N} \subset \mathbb{Z} \subset \mathbb{Q} \subset \mathbb{R}.$$

The arithmetical operations in $\mathbb{R}$ can be defined, for instance, as the limit of operations on rationals, but we shall not be concerned with this problem here. All we need to note is that, like in $\mathbb{Q}$, in $\mathbb{R}$ we can perform all four arithmetical operations unrestrictedly, excluding division by zero. So $\mathbb{R}$, like $\mathbb{Q}$, is a *field.*

A distance in $\mathbb{R}$

We define a distance in $\mathbb{R}$ in the same way as we did in $\mathbb{Q}$. The distance between two real numbers x and y is defined as the size of the real number $x - y$, which is $|x - y|$. If two real numbers share the beginning of their sequence of digits, then they are close to each other: the larger the number of common digits, the closer they are. For instance, let us consider the three numbers

$$\begin{aligned}
\pi &= 3.14159265358979\ldots \\
r_1 &= 3.14158430114199\ldots \\
r_2 &= 3.14159265359001\ldots
\end{aligned}$$

We see that r_1 has 4 digits after the decimal point in common with π, while r_2 has 9 of them. We find

$$
\begin{aligned}
|\pi - r_1| &= 0.000008\ldots \simeq 10^{-5} \\
|\pi - r_2| &= 0.00000000008\ldots \simeq 10^{-10}
\end{aligned}
$$

The converse is not true: two real numbers may be very close to each other without sharing any digit at all. This phenomenon originates from the fact that certain rationals have two distinct decimal representations (cf. Theorem 3.2). If x is one such rational, then closeness to x means identification of the digits of *one of the two* representations, that is

$$
\begin{aligned}
x &= \tfrac{1}{2} = 0.500000000000000\ldots = 0.499999999999999\ldots \\
r_1 &= 0.50008430114199\ldots \\
r_2 &= 0.49999999999001\ldots
\end{aligned}
$$

Thus, r_2 is closer to $1/2$ than r_1.

Real numbers in Maple

A real number in Maple is a decimal number with radix point and sign. The positive sign may be omitted.

```
> 1.5, -33.0, 0.0001, +41.41;
```

$$1.5, -33.0, .0001, 41.41$$

```
> seq(whattype(x),x=%);
```

$$float, \, float, \, float, \, float$$

The above syntactical construction with **seq** is novel, since the running index x is extracted from the elements of an arbitrary expression sequence, rather than from a range. The expression sequence can also appear directly as the second argument of **seq**, but in this case it must be enclosed in parentheses, because a call to **seq** requires two arguments.

```
> seq(whattype(x),x=(1.5,-33.0,0.0001,+41.41));
```

We shall return to this point in section 6.3.

The word *float* denotes the *floating-point* data type, which refers to the way these data are represented in Maple (see below).

Very large or very small real numbers are more conveniently written in *exponential* notation, which mimics more closely the way these numbers are represented. The syntax is the following

$$x\mathrm{e}y \longrightarrow x10^y$$

where x is a real or an integer, y is an integer, and they are separated by the character **e**. This corresponds to multiplying x by 10^y. Thus, `1.5e-4` represents the number $1.5 \cdot 10^{-4} = 0.00015$.

```
> 1.5e-4,1.5e4,1.5e-20,1.5e20;
```

$$.00015, 15000., .15 \, 10^{-19}, .15 \, 10^{21}$$

The function `evalf` can be used to transform integer or rational data types (among others) into the floating-point type

```
> 3,1/3,evalf(3),evalf(1/3);
```

$$3, \frac{1}{3}, 3., .3333333333$$

Expressions such as `3.` and `.3` are valid real numbers, but their use is discouraged because the radix point is easily overlooked, as illustrated in the following example

```
> (1+2)/(3+4)*5+6-7/8;
```

$$\frac{407}{56}$$

```
> (1+2)/(3+4)*5+6-7/8.;
```

$$7.267857143$$

Note that the floating-point data type is 'contagious'. The presence of the single real number `8.` forces the whole expression to become real.

The floating-point representation is non-exact, since a computer cannot store infinitely many digits. It should be noted that while integers and rationals can also be represented as floating-point numbers, the two representations are vastly different. Every floating-point number is represented as a *pair of integers*, called the *mantissa* and the *exponent*.

$$3.141592654 = mantissa \cdot 10^{\,exponent} = 3141592654 \cdot 10^{-9}$$

Mantissa and exponent are *operands* (that is, components) of the expression representing the real number

```
> op(evalf(Pi));
```

$$3141592654, -9$$

The function `op` returns the operands of an expression (see chapter 6).

We compare the representations of $1/3$ as a rational and as a real, with 10 digits accuracy

```
> op(1/3);op(evalf(1/3));
```

$$1, 3$$
$$3333333333, -10$$

This result says that

$$\frac{1}{3} \simeq 3333333333 \cdot 10^{-10}$$

We stress again that the floating-point representation is only approximate

```
> 3*(1/3),3*evalf(1/3);
```

$$1, .9999999999$$

Exercises

Exercise 5.5. The following numbers are rational approximations to $\pi = 3.1415\ldots$

$$\frac{22}{7}, \quad \frac{314}{100}, \quad \frac{333}{106}, \quad \frac{355}{113}, \quad \frac{355}{114}, \quad \frac{103993}{33102}, \quad \frac{31415926}{10000000}. \qquad (5.8)$$

(a) Create a Maple *list* whose elements are these numbers.

(b) Define a Maple function d(x), which returns the distance between x and π: $d(x) = |x - \pi|$.

(c) Compute the distance between π and every element of the sequence (5.8), whence order the elements of the sequence (5.8) according to how well they approximate π (begin with the worst approximation).

Exercise 5.6. Let α be a real number, and let

$$f_\alpha(x) = \alpha x - \frac{1}{2} \qquad g(x) = e^{-(x-2)^6}.$$

For certain values of α, the equation $f_\alpha(x) = g(x)$ admits three solutions, which correspond to three intersections of the graphs of f_α and g. These values of α form an interval $[a, b]$. Plot f_α and g in the same plot, for various values of α of your choice, and by examining the graphic data determine approximate values for a and b (10% accuracy will suffice).

Exercise 5.7. Let D be a positive integer. We consider the following recursive sequence of rational numbers

$$x_0 = 1 \qquad x_{t+1} = \frac{x_t^2 + D}{2\,x_t}, \qquad t \geq 0.$$

It can be shown that the elements of this sequence approach $\sqrt{D}$ very quickly. This method for computing the square root is called *Newton's method*.

Using the above sequence twice, determine a rational number approximating $\sqrt{2} + \sqrt{3}$ with an error not greater than 10^{-5}.

Exercise 5.8*. Prove that any decimal expansion with finitely many digits to the left of the radix point defines a unique number (you must prove the convergence of a series).

5.3 Random and pseudo-random digits*

The examples of irrational numbers that were given in (5.7) are, in a sense, misleading. There are obvious *patterns* in their digit sequences. The knowledge of such patterns can be exploited to specify these numbers concisely, using a very small amount of information. For instance, the number α_1 is the real number whose digits are given by the sequence of all positive integers. Likewise, the number α_2 is defined by the rule that its nth digit is equal to 4 if n is a number of the form $n = k(k + 1)/2$ (a *triangular number*), and to 0 if it is not, etc. These definitions could be translated into computer programs that generate these numbers to any desired degree of accuracy.

However, the digits of a real number are not subject to any constraint, in the sense that *any* sequence of digits defines a unique number. So, if we choose a number blindly, we should expect its digit sequence to be not only aperiodic, but also patternless. The validity of this observation is confirmed by a celebrated theorem of complexity theory, which states that *almost all* numbers have *random* digits.

The expression 'almost all' is to be intended in a probabilistic sense: a blind choice will give you a random number unless you are extremely lucky (or extremely unlucky, depending upon which way you look at it). Said differently, if you take all possible strings of n digits, where n is a large integer, those that are not random form a vanishing fraction of the total.

Randomness is the property we usually associate to the outcome of a gambling device. Loosely speaking, a sequence of digits is said to be 'random' if the most economical way of specifying its elements is to list them all explicitly, one by one. So a sequence is random if the information it contains cannot be compressed in any way. Any computer program that generates the first n bits of a random sequence cannot itself be shorter than n bits, lest we would have achieved a compression of information stored in the sequence.

To illustrate this concept, suppose we have to send to a distant galaxy the rather dull sequence consisting of the first billion odd integers, and suppose that the cost of the transmission is very high. It is quite possible to minimize the amount of transmitted data: instead of sending the entire sequence, we send the Maple statement `seq(2*i-1,i=1..10^9);`. This is a compressed version of the same information (21 characters in all!), which the receiver can decode, recovering the original information. This substantial data compression was made possible by the very non-random nature of our sequence. But if we had to transmit, say, the outcome of all football games since the beginning of the century, one can see that it is going to cost a lot more.

We seem to have reached an impasse. If we are to build a very long string of digits, then we need some kind of *algorithm* — a sequence of computer instructions to be repeated automatically. However, the very existence of such algorithm makes the outcome non-random, according to our definition of randomness. It then appears that a truly random sequence cannot be meaningfully defined, represented, computed, or predicted: it is terminally beyond our grasp. But then, how close can one get to observing randomness inside a computer? What a machine can generate are the so-called *pseudo-random numbers:* they 'look like' random numbers, but they are actually produced by programs which are often small and fast.

To see how this could be achieved one does not have to go too far, because pseudo-randomness is ubiquitous in arithmetic. Even the digits of a rational number *before they start repeating* already possess a considerable degree of unpredictability. For instance, the repeating block of digits of any rational with prime denominator 47 happens to be 46 digits long. By varying the numerator, we have an easy method for constructing finite pseudo-random sequences

```
> evalf(6/47,46);evalf(13/47,46);evalf(36/47,46);
```

$$.1276595744680851063829787234042553191489361702$$

$$.2765957446808510638297872340425531914893617021$$

$$.7659574468085106382978723404255319148936170213$$

We see that the above sequences are cycling permutations of the same basic sequence, with the numerator determining the starting point. A very popular type of pseudo-random number generators employs a variant of the above principle (check Maple's facilities on random number generators by typing ?rand).

An even more dramatic illustration is offered by the digits of the square root of two. Because $\sqrt{2}$ is irrational (see below), its digits never repeat. We display the first 50 of them

```
> sqrt(2)=evalf(sqrt(2),50);
```

$$\sqrt{2} = 1.4142135623730950488016887242096980785696718753769$$

and note that they have a random flavour. However, this number is defined by a tiny amount of information: it is one of the two solutions of the polynomial equation $x^2 = 2$, the other being $-\sqrt{2}$ (incidentally, that's how Maple represents it). Therefore, according to our previous definition, we are not allowed to call these digits random. Does it mean that some pattern *must* exist? None has been found so far, and, moreover, there are several phenomena related to the digits of $\sqrt{2}$ which carry the signature of randomness. Is our definition of randomness appropriate then? The fact is that the question *'how random are the digits of $\sqrt{2}$?'* remains unanswered, and so do many deep problems surrounding pseudo-randomness. For an introduction to the modern notions of deterministic randomness and chaos, see Reference [6].

We conclude this brief exploration of $\sqrt{2}$ by proving that it is irrational, which implies that its digits do not repeat. Establishing irrationality of a given real number is usually very difficult, but in this case we can prove it without much effort.

Theorem 5 $\sqrt{2} \notin \mathbb{Q}$.

Proof. Assume $\sqrt{2}$ is rational. Then there exist two integers a and b such that $\sqrt{2} = a/b$, and we shall agree that a and b are relatively prime (if they are not, we first divide by their greatest common divisor). Squaring, we obtain $2b^2 = a^2$, whence a^2 is even. From the fundamental theorem of arithmetic, we can then conclude that a is even (why?), and we let $a = 2k$. But then $2b^2 = (2k)^2 = 4k^2$ and so $b^2 = 2k^2$, and with the same reasoning we conclude that also b is even. But this is impossible, because a and b were assumed to have no common divisor. So $\sqrt{2}$ is irrational.

This is an example of *proof by contradiction*. We have assumed the opposite of what we wanted to prove, and we have arrived at an internal inconsistency in the argument

$$\sqrt{2} \text{ is rational} \quad \Longrightarrow \quad \text{two relatively prime integers are even.}$$

This necessarily means that the statement on the left of the implication $\Longrightarrow$ is false, that is, that its negation is true.

5.4   Complex numbers

There are still arithmetical problems in $\mathbb{R}$, and they originate from the fact that the square of a real number cannot be negative. Suppose that this was not true, and that we could find a real number y and a *positive* real number a such that $y^2 = -a$. Then by writing $a = \sqrt{a}\sqrt{a}$, and letting $x = y/\sqrt{a}$, we would have a *real* number x (why real?) satisfying the equation $x^2 = -1$, a nice quadratic equation with integer coefficients. So all problems in $\mathbb{R}$ can be traced to the impossibility of solving the equation $x^2 = -1$, for a real number x. We have experienced similar problems in $\mathbb{Q}$, when we attempted to solve equations like $x^2 = 2$. That led to the introduction of $\mathbb{R}$.

A solution of the equation $x^2 = -1$ is '*a thing whose square is equal to minus one*', which is often called the *imaginary unit* and denoted by the letter i. We shall not attempt to ascertain what such object actually *is;* all we need to know is that if we square it, we get -1. If we assume that i obeys ordinary arithmetical rules, we must then conclude that $-i$ also satisfies the same equation, since $(-i)^2 = (-1)^2 i^2 = 1 \cdot i^2 = -1$. We have

$$i^0 = 1, \quad i^1 = i, \quad i^2 = -1, \quad i^3 = -i, \quad i^4 = 1, \quad \text{etc.} \quad (5.9)$$

That is, the sequence of powers of i, namely $i^k, k = 0, 1, 2, \ldots$ is *periodic* with period 4. Maple represents $\sqrt{-1}$ by the symbol I and we can easily generate a few terms of this sequence, to emphasize its periodicity

```
> seq(I^k,k=0..12);
```

$$1, I, -1, -I, 1, I, -1, -I, 1, I, -1 - I, 1$$

This is not the first time we find a number whose powers form a periodic sequence. The powers of the integers 1 and -1 also generate periodic sequences, with period 1 and 2, respectively

```
> seq((-1)^k,k=0..12);
```

$$1, -1, 1, -1, 1, -1, 1, -1, 1, -1, 1 - 1, 1$$

So far we have considered i in isolation. Now we combine it with elements of $\mathbb{R}$, beginning with expressions of the type

$$z = r_0 + r_1 i + r_2 i^2 + r_3 i^3 + r_4 i^4 + \cdots + r_n i^n \qquad r_k \in \mathbb{R}.$$

Because of (5.9), this expression becomes

$$z = r_0 + r_1 i - r_2 - r_3 i + r_4 + r_5 i + \cdots = (r_0 - r_2 + r_4 - \cdots) + i(r_1 - r_3 + r_5 - \cdots).$$

The term in the first set of parentheses is a *finite* sum of real numbers, and it is therefore a real number. (Infinite sums may also be meaningful, but their study transcends the scope of this course.) The same holds for the sum in the second set of parentheses, and so z is of the form

$$z = r + is \qquad r, s \in \mathbb{R}. \tag{5.10}$$

The real numbers r and s are called the *real part* and *imaginary part* of z, respectively.

Next we consider the four arithmetical operations with expressions of the type (5.10). Let $z = r + is$ and $w = t + iu$. We have

$$
\begin{aligned}
z \pm w &= r + is \pm t \pm iu = (r \pm t) + i(s \pm u) \\
zw &= (r + is)(t + iu) = (rt - su) + i(ru + ts) \\
\frac{z}{w} &= \frac{r + is}{t + iu} = \frac{(r + is)(t - iu)}{(t + iu)(t - iu)} = \frac{(rt + su) + i(st - ru)}{t^2 + u^2} \qquad (5.11) \\
&= \frac{rt + su}{t^2 + u^2} + i\frac{st - ru}{t^2 + u^2}.
\end{aligned}
$$

For division to make sense, we must assume that all denominators are nonzero. For this it is sufficient to require that $w \neq 0$, that is, that t and u are not both zero. This in turn guarantees that the quantity $t - iu$ is also nonzero, and so is the product $(t + iu)(t - iu) = t^2 + u^2$.

Equations (5.11) show that all numbers of the form (5.10) form a set closed under the four arithmetical operations, excluding division by zero. This is the set $\mathbb{C}$ of *complex numbers*. So $\mathbb{C}$ is a *field*. We see that $\mathbb{C}$ is obtained from $\mathbb{R}$ by *adjoining* to it the quantity i, and then allowing unrestricted arithmetical operations between i and the elements of $\mathbb{R}$, assuming that i obeys ordinary arithmetical rules.

Letting $s = 0$ in (5.11), we obtain $z = r + 0 \cdot i = r$, a real number. Thus, $\mathbb{C}$ contains $\mathbb{R}$, the set of complex numbers with zero imaginary part, and we have the chain of inclusions

$$\mathbb{N} \subset \mathbb{Z} \subset \mathbb{Q} \subset \mathbb{R} \subset \mathbb{C}. \tag{5.12}$$

As far as 'ordinary' numbers are concerned, this is the end of the road. There is no arithmetical need of going beyond $\mathbb{C}$. The distinguished 19th-century mathematician L. Kronecker once said:

'God gave us the integers, the rest is the work of man.'

You can begin to see the giant edifice mathematicians have constructed from the natural numbers alone. The difficult step is the one between $\mathbb{Q}$ and $\mathbb{R}$. In fact, only two integers are needed to construct a rational, and only two reals are needed to construct a complex number. But *infinitely many* rationals are needed to construct a real number. In fact, the domain lying between $\mathbb{Q}$ and $\mathbb{R}$ (or $\mathbb{C}$) is by far the most interesting one, but its study lies beyond the scope of this course.

The following table summarizes the most significant steps in the journey from $\mathbb{N}$ to $\mathbb{C}$. The need of enlarging a set of numbers originates from the impossibility of solving certain equations.

FROM $\mathbb{N}$ TO $\mathbb{C}$

set	solve $x + a = b$ $a, b \in \mathbb{N}$	solve $ax = b$ $a, b \in \mathbb{Z}, a \neq 0$	solve $x^2 = a$ $a \in \mathbb{N}$	solve $x^2 = a$ $a \in \mathbb{Z}$
$\mathbb{N}$	no	no	no	no
$\mathbb{Z}$	yes	no	no	no
$\mathbb{Q}$	yes	yes	no	no
$\mathbb{R}$	yes	yes	yes	no
$\mathbb{C}$	yes	yes	yes	yes

A distance in $\mathbb{C}$

How do we measure the *size* of a complex number? The real numbers were represented geometrically as points on a line, and we defined their size to be the distance from the origin. Likewise, complex numbers are points in the plane, so it is natural to represent the size of a number as its distance from the origin in $\mathbb{C}$.

If $z = x + iy$, we let $\bar{z} = x - iy$. The number $\bar{z}$ is called the *complex conjugate* of z, and from (5.11) we find

$$z\bar{z} = (x + iy)(x - iy) = x^2 + y^2.$$

But then, from Pythagoras' theorem, the distance of the point z from the origin is given by the *modulus* of z

$$|z| = \sqrt{z\bar{z}} = \sqrt{x^2 + y^2}. \tag{5.13}$$

Since

$$\overline{(\bar{z})} = \overline{x - iy} = x + iy = z,$$

we have that
$$|\bar{z}| = \sqrt{\bar{z}\bar{\bar{z}}} = \sqrt{\bar{z}z} = |z|.$$

As we did for $\mathbb{Q}$ and then for $\mathbb{R}$, we define the distance between two complex numbers z_1 and z_2 as the size of their difference, which is $|z_1 - z_2|$. On the plane, the triangle inequality (2.6) justifies its name: it says that if x, y, and z are complex numbers representing the vertices of a triangle, then the length of the side connecting x to z cannot be greater than the sum of the length of the other two sides.

So in $\mathbb{C}$, like in $\mathbb{R}$, there is an arithmetical structure and a distance (but *not* an ordering relation $\leq$).

Complex numbers in Maple

We have seen that in Maple the number $\sqrt{-1}$ is represented by the symbolic name I. The basic operations of complex arithmetic are straightforward, with several intrinsic functions available for this purpose.

```
> z:=2-I:
> Re(z),Im(z),conjugate(z),abs(z);
```
$$2, -1, 2 + I, \sqrt{5}$$

An expression involving complex numbers with *rational* real and imaginary parts is automatically simplified to the form $x + iy$ with $x, y \in \mathbb{Q}$

```
> z^2+1/z;
```
$$\frac{17}{5} - \frac{19}{5} I$$

```
> evalf(%);
```
$$3.400000000 - 3.800000000\, I$$

Note that `evalf` acts separately on the real and imaginary parts of a complex expression (see also section 6.3).

The simplification to the form $r + is$ is not carried out in general. To achieve it, one must use the standard library function `evalc` (evaluate to complex)

```
> 1/(x+I*y);
```
$$\frac{1}{x + I\,y}$$

```
> evalc(%);
```
$$\frac{x}{x^2 + y^2} - \frac{I\,y}{x^2 + y^2}$$

Example 5.1. *Plotting a complex sequence.* We want to plot the complex numbers
$$z_n = \frac{1 - n\,i}{n + 4\,i^n} \qquad n = 1, 2, \ldots$$

as points in the complex plane. The data are arranged as a list consisting of two-element lists $[x_n, y_n]$, where x_n and y_n are the real and imaginary part of z_n, respectively.

```
> z:=n->(1-n*I)/(n+4*I^n):
> data:=[seq([Re(z(n)),Im(z(n))],n=1..30)]:
> plot(data);
```

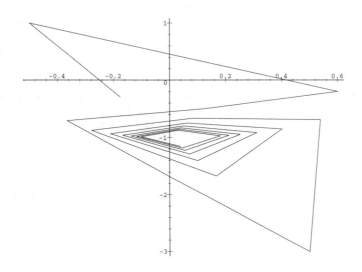

Exercises

Exercise 5.9.   Let $z = 11 - 19i$. Compute, with Maple

$$1/z, \quad \bar{z}, \quad |z|, \quad \arg(z), \quad \text{Re}(z), \quad \text{Im}(\bar{z}), \quad \overline{1 - \bar{z}^2}.$$

where $\bar{z}$ is the complex conjugate of z.

Exercise 5.10.   Which of the following complex numbers

$$z_1 = \frac{109}{11} \qquad z_2 = 7 + 7i \qquad z_3 = -5 - 9i \qquad z_4 = 1 - 10i$$

is nearest to the origin? Which is nearest to $-i$?

Exercise 5.11.   Consider the following recursive sequence of complex numbers

$$z_0 = -\frac{4}{41}(9 + i) \qquad z_{t+1} = 8\frac{z_t - i}{i\,z_t + 10}, \qquad t \geq 0.$$

(a) Show that z_1, z_2, and z_3 lie on a circle centered at the point $z = 4 + i$, and determine its radius.

(b) Let z_t be as above. Determine the smallest value of t for which z_t lies inside a circle of diameter $1/10$ centered at $-2i$.

Exercise 5.12. Let n be a fixed positive integer. Consider the following sequence of complex numbers

$$z_k \;=\; \exp(2\pi i k/n) \qquad\qquad k \geq 0.$$

(a) Construct a Maple function `z(k,n)` whose value is z_k, hence display z_k in the form $x_k + iy_k$.

(b) Prove with Maple that z_k lie on the *unit circle* in the complex plane (the circle with radius 1 and centre at $z = 0$).

(c) Let $n = 7$. Express z_1 in the form $x_1 + iy_1$, first symbolically, and then numerically, with 20 digits of accuracy.

(d) If $z = x + iy$, from (5.13) we have that

$$|z|^2 \;=\; x^2 + y^2.$$

Using the above equation and the numerical expression for z_1 just computed, verify that $|z_1|^2 \;=\; 1$.

(e) Fix $n = 5$. Plot the points $z_1, z_2, \ldots, z_5$ in the complex plane, connecting successive points by segments.

(f) Plot a regular polygon with 11 sides.

(g) For $n = 11$, generate and plot the sequence of points

$$w_k \;=\; z_{3k} \qquad\qquad k = 0, 1, \ldots, 11.$$

Identify w_4 on your plot.

(h) By now you should have enough evidence to conjecture that for any $n > 0$ the sequence $k \mapsto z_k$ is *periodic* with period n. This means that for every integer m, the following holds

$$z_k \;=\; z_{k+mn} \qquad\qquad k \geq 0.$$

Prove it.

Exercise 5.13. Let the complex numbers $z_1 = -91 - 24i$ and $z_2 = 200 + 100i$ be two vertices of a square in the complex plane. There are three such squares (think about it).

(a) Determine the areas of the three squares.

(b) Construct the three squares and plot them, all in the same plot. (Maple may scale the two axes differently, resulting in a distorted picture. To prevent this from happening, insert the option `scaling=constrained` as the last argument of the function `plot`.)

Exercise 5.14. Prove de Moivre's theorem with Maple

$$\left(e^{i\theta}\right)^n = \cos(n\theta) + i\sin(n\theta).$$

5.5   Standard library functions

Maple has several built-in functions for performing the basic operations of real and complex calculus. The argument of the trigonometric functions is in radians.

REAL AND COMPLEX FUNCTIONS

Trigonometric, hyperbolic, and logarithmic functions

sin(x),cos(x),tan(x)	Sine, cosine, tangent of the angle in radians.
sec(x),csc(x),cot(x)	Secant, cosecant, cotangent.
arcsin(x),arccos(x), arctan(x)	Arc sine, arc cosine, arc tangent, with range in $[-\pi/2, \pi/2]$.
sinh(x),cosh(x),tanh(x)	Hyperbolic sine, cosine, tangent.
sech(x),csch(x),coth(x)	Hyperbolic secant, cosecant, cotangent.
sqrt(x)	Square root.
exp(x)	Exponential function e^x.
log(x)	Natural logarithm (base e).
log10(x)	Logarithm to the base 10.
log[b](x)	Logarithm to the base b.
GAMMA(x)	Gamma function.

Other functions

abs(x)	Absolute value.
sign(x)	Sign of x (1 or -1).
max(x,y,...)	The largest value among arguments.
min(x,y,...)	The smallest value among arguments.
round(x)	Round x to the nearest integer.
trunc(x)	Truncate x to its integer part.
floor(x)	Floor function $\lfloor x \rfloor$.
ceil(x)	Ceiling function $\lceil x \rceil$.
evalf(z,n)	Evaluate z to n floating-point digits.

Complex arithmetic

Re(z)	Real part of z.
Im(z)	Imaginary part of z.
conjugate(z)	Complex conjugate of z.
argument(z)	Argument of z (principal value).
evalc(z)	Split z into real and imaginary part.

Differentiation and Integration

diff(expr,x)   Derivative of expr with respect to x.
int(expr,x)   Indefinite integral of expr with respect to x.

Example 5.2. Let x be any real expression, whose value is nonzero. The function

```
> ndgt:=x->max(0,floor(evalf(log[10](abs(x)))) + 1):
```

returns the number of decimal digits of the integer part of x.

Example 5.3. Let x be a rational number. Then the expression iquo(numer(x)),denom(x)) has the same value as floor(x) if x is non-negative, and as ceil(x), if x is negative.

Example 5.4. The factorial function is well-approximated by the following formula

$$n! = \left(\frac{n}{e}\right)^n \sqrt{2\pi n} \qquad (5.14)$$

called *Stirling's formula*. Here $e = 2.718\ldots$ is *Napier's constant* (the basis of the natural logarithm), which can be computed with Maple using the exponential function

```
> evalf(exp(1),50);
```

$$2.7182818284590452353602874713526624977572470937000$$

The accuracy of Stirling's formula increases as n increases, in the sense that the *relative error* vanishes. This means that the *ratio* between $n!$ and Stirling's approximation approaches unity.

```
> Stirling:=n->evalf((n/exp(1))^n*sqrt(2*Pi*n)):
> 10!/Stirling(10),100!/Stirling(100);
```

$$1.008365358, \ 1.000833670$$

Note that the *absolute error* — the size of the *difference* between the two quantities — increases indefinitely!

```
> 10!-Stirling(10),100!-Stirling(100);
```

$$30104.378, \ .7773848^{155}$$

The gamma function

The gamma function $\Gamma(x)$ is defined as the definite integral

$$\Gamma(x) = \int_0^\infty t^{x-1} e^{-t} dt, \qquad x > 0. \qquad (5.15)$$

The Γ function may be viewed as a generalization of the factorial function. Specifically, we now show that if x is a non-negative integer, then $\Gamma(x+1) = x!$.

We replace x by $x + 1$ in (5.15) and integrate by parts, recalling that $\int e^{-t} dt = -e^{-t}$.

$$
\begin{aligned}
\Gamma(x + 1) &= t^x \left(-e^{-t} \right)\big|_0^\infty + \int_0^\infty x t^{x-1} e^{-t} dt \\
&= x \int_0^\infty t^{x-1} e^{-t} dt \\
&= x \, \Gamma(x).
\end{aligned}
$$

Thus, $\Gamma(x + 1) = x \, \Gamma(x)$. If we let $\Gamma(x + 1) = x!$, we obtain $x! = x \cdot (x - 1)!$, which is the recursion relation for the factorial function. To prove that $\Gamma(x + 1)$ is indeed equal to $x!$, we must check the initial condition $0! = 1 = \Gamma(1)$

$$
\Gamma(1) = \int_0^\infty t^0 e^{-t} dt = \int_0^\infty e^{-t} dt = \left(-e^{-t} \right)\big|_0^\infty = 1.
$$

This shows that $\Gamma(x + 1) = x!$. Note, however, that the integral (5.15) is defined also when x is not an integer.

In Maple, the symbolic name for the gamma function is GAMMA. Maple knows that the value of Γ at positive integers is an integer. It also knows the values of Γ at half-integers.

```
> seq(GAMMA(x),x=1..5);
```

$$1, 1, 2, 6, 24$$

```
> seq(GAMMA(evalf(x)),x=1..5);
```

$$1., 1., 2., 6., 24.$$

```
> seq(GAMMA(x-1/2),x=1..5);
```

$$\sqrt{\pi}, \; \frac{1}{2} \sqrt{\pi}, \; \frac{3}{4} \sqrt{\pi}, \; \frac{15}{8} \sqrt{\pi}, \; \frac{105}{16} \sqrt{\pi}$$

```
> seq(GAMMA(evalf(x-1/2)),x=1..5);
```

$$1.772453851, .8862269255, 1.329340388, 3.323350970, 11.63172840$$

When handling expressions involving factorials, Maple often converts factorials to gamma functions.

```
> n!-1/(n+3)!;
```

$$n! - \frac{1}{(n + 3)!}$$

```
> simplify(%);
```

$$\frac{\Gamma(n + 1)\,\Gamma(n + 4) - 1}{\Gamma(n + 4)}$$

Plotting functions of a real variable

In chapter 4, we introduced the utility function `plot`, to plot the elements of a sequence. They were represented as a *discrete* set of points in the plane, implemented in Maple as a list consisting of two-element lists containing the Cartesian coordinates of each point.

Plotting a function of a real variable involves simulating a *continuum* of values, which is done by a fine scanning of the domain of the function. The function `plot` will now require specifying the function to be plotted and its domain, as the first and second argument, respectively. The co-domain is then adjusted automatically, and so is the frequency with which the domain is scanned.

In the following example, we plot a function displaying rapid oscillations

```
> plot(sin(x)*sin(10*x)*sin(100*x),x=0..Pi);
```

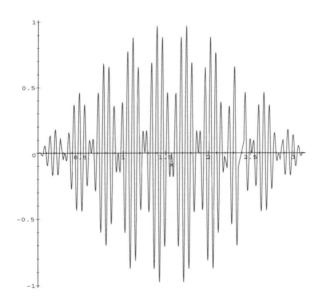

To gain information on the detailed structure of our function, we resort to magnification. This is achieved by a suitable compression of the domain and the co-domain of the function, the latter entered as an additional argument of `plot`

```
> plot(sin(x)*sin(10*x)*sin(100*x),x=0..0.5,y=-0.5..0.5);
```

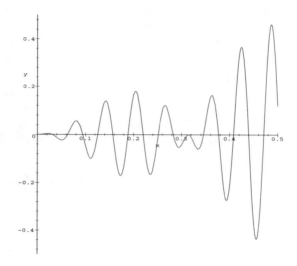

Chapter 6

Structure of expressions

In previous chapters, we have encountered a variety of data types, e.g.,

```
> 3,2/3,2.95,joe,[a,b],{a,b}:
> seq(whattype(x),x=%),whattype(%);
```

$$integer, fraction, float, symbol, list, set, exprseq$$

In this chapter, we consider the way Maple represents *composite data,* of which sets and lists are simple examples. In particular, we analyze in detail the sum, product, and exponent data types, which are used to represent algebraic expressions. This chapter is more technical than mathematical, but it is necessary for the arithmetic of polynomials and rational functions, and for understanding the behaviour of certain functions acting on expressions, such as `map`, `seq`, `select`, etc.

6.1 Analysis of an expression

How does Maple represent an expression such as $a + 2bc - b^3$?

```
> x:=a+2*b*c-b^3;
```

$$x := a + 2bc - b^3 \qquad (6.1)$$

```
> whattype(x);
```

$$+$$

In the first instance, Maple regards x as a datum of the *sum* type. To list the *operands* of this expression, we use the function `op`

```
> op(x);
```

$$a, \; 2bc, \; -b^3$$

```
> whattype(%);
```

$$exprseq$$

Thus, x is represented as the sum of the three operands a, $2bc$, and $-b^3$. Graphically

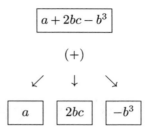

Figure 6.1: Graphical representation of the expression $a + 2bc - b^3$: it is of type + and has three operands.

The standard library function op is a general-purpose function for the analysis of an expression. The command
```
> op(expression);
```
returns the operands of an expression in the form of an expression sequence. The function op has a first optional argument. The command
```
> op(n, expression);
```
returns the nth element in the sequence of operands of *expression*.
```
> op(2,x);
```

$$2bc$$

The first argument of op can also be a *range*
```
> op(2..3,x);
```

$$2bc, -b^3$$

We continue analyzing expression (6.1), by looking individually at its operands
```
> left:=op(1,x):
> middle:=op(2,x):
> right:=op(3,x):
```
The leftmost operand a has no inner structure — its data type is *primitive*
```
> left,op(left),whattype(left);
```

$$a, a, symbol$$

and its analysis is complete. The middle operand $2bc$ is an expression of the *product* type, consisting of three operands
```
> whattype(middle);
```

$$*$$

```
> op(middle);
```

$$2, b, c$$

These operands are primitive

```
> seq(whattype(z),z=%);
```

integer, symbol, symbol

The rightmost operand is also a product

```
> whattype(right);
```

$$*$$

```
> op(right);
```

$$-1, b^3$$

```
> seq(whattype(z),z=%);
```

integer, ^

The expression b^3 is an exponent, with operands b and 3

```
> op(op(2,[%%]));
```

$$b, 3$$

(The use of square brackets is necessary here, lest the innermost op would have been called with the wrong number of arguments.) The above sub-expressions are represented as

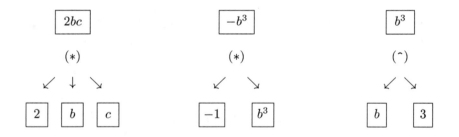

Figure 6.2: Representations of the sub-expression $2bc$, $-b^3$, and b^3, respectively. Note that b^3 is a sub-expression of $-b^3$.

The analysis of the expression (6.1) is now complete. Assembling the figures (6.1) and (6.2), we obtain the structure depicted in figure 6.3, which is called a *tree.*

For some composite data types (e.g., lists, sets), the *selection operator* [   ] may be used as a variant of op, with a slightly different behaviour

```
> L:=[one,two,three]:
> L[3],op(3,L),L[1..2],op(1..2,L),L[1..1];
```

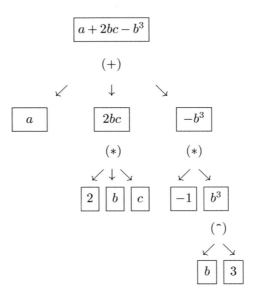

Figure 6.3: The tree associated with the complete representation of the expression $a + 2bc - b^3$.

$$three,\ three,\ [one,\ two],\ one,\ two,\ [one]$$

We see that if the selection is performed by specifying a *range,* the selection operator will reassemble the selected operand into the original data type. In addition, the selection operator may be used to assign a new value to an element of the list

```
> L[2]:=TWO:
> L;
```

$$[\,one,\ TWO,\ three\,]$$

However, direct assignment to elements of a list is not efficient and should be avoided. To modify a list, it is preferable to construct a *copy* of the original list, using `subsop` (see section 6.2 and online documentation on `subsop` and `selection`).

The standard library function `length` (see section 2.7), when applied to an arbitrary expression, will compute recursively the length of each operand, and then add up the result. This quantifies the size of the expression, although the result is not always obvious

```
> length(12345),length(12345*x),length(x->x^2);
```

$$5,\ 9,\ 28$$

6.2 More on substitutions

The standard library function `subs` substitutes expressions with other expressions. The expression to be changed must appear as an operand of the original expression or of one of its sub-expressions. The following examples illustrate how `subs` operates, depending on the structure of the target expression.

```
> expr1:=x*y+z:
> expr2:=x*y*z:
> whattype(expr1),whattype(expr2);
```

$$+, *$$

```
> op(expr1);op(expr2);
```

$$xy, z$$
$$x, y, z$$

Note that xy is a sub-expression of `expr1` but not of `expr2`.

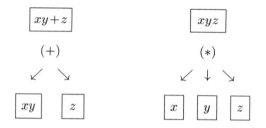

Figure 6.4: Representation of $xy + z$ and xyz. Only $xy + z$ contains xy as a sub-expression.

Hence

```
> subs(x*y=joe, expr1);
```

$$joe + z$$

```
> subs(x*y=joe,expr2);
```

$$x\,y\,z$$

To replace xy in `expr2`, we must resort to a trick

```
> subs(x=joe/y, expr2);
```

$$joe\,z$$

We consider another example

```
> beta:=a*b*c+b*c+(b*c)^3;
```

$$\beta := abc + bc + b^3 c^3$$

```
> op(beta);
```

$$abc,\ bc,\ b^3\,c^3$$

Note that the expression bc does not appear as a sub-expression of the third operand, due to simplification.

```
> subs(b=y,beta);
```

$$ayc + yc + y^3c^3$$

```
> subs(b*c=y,beta);
```

$$abc + y + b^3c^3$$

In order to target the substitution to a particular operand of an expression, we use the standard library function `subsop`, which substitutes specified operands with new values

```
> subsop(3=three, beta);
```

$$abc + bc + three$$

The data type *fraction* is reserved for rationals. An algebraic fraction is instead regarded as a *product*

```
> 2/3:whattype(%),op(%);
```

$$fraction,\ 2,\ 3$$

```
> a/b:whattype(%),op(%);
```

$$*,\ a,\ \frac{1}{b}$$

```
> op(3,[%]):whattype(%),op(%);
```

$$\char`^,\ b,\ -1$$

Thus, a/b is represented as $a \cdot b^{-1}$.

Exercises

Exercise 6.1. Construct the tree representation of the following algebraic expressions

(1) $\dfrac{2}{y^2}$

(2) $x^y - \dfrac{z}{w}$

(3) $x^{yz} + w$

(4) $\sqrt{xy - xz + yzw}$

(5) $\dfrac{x - y}{\sqrt{z + \sqrt{w}}}$

(6) $\dfrac{x}{y + \dfrac{z}{w}}.$

Exercise 6.2. Construct the tree representation of the following Maple expressions

(1) `[2,3,{2*3}]`

(2) `seq(i+x,i=1..n)`

(3) `33^22<22^33`

(4) `[x,[y],[[z]]]`

(5) `{[[],([])],{}}`

(6) `igcd(22,igcd(a,b))`

(7) `A union (B intersect C)`

(8) `x->{op(1,x)}`

Exercise 6.3*. Explore the tree representation of a factored integer, designed to prevent automatic expansion of the factors.

6.3 Functions acting on operands of expressions

Several standard library functions make implicit use of the operand function op, by acting in various ways on the operands of an expression. Here we consider seq, map, select, remove, evalf. Other functions acting on operands (add, mul, sum, prod) will be dealt with in chapter 8.

In chapter 4, we learned that the second argument of the function seq is either a range, which is expanded to an expression sequence, or an expression sequence itself. Such argument, however, can be virtually any expression, which Maple turns into the expression sequence consisting of its operands at the top level. This feature makes seq into one of the most useful utility functions in Maple. For instance, the following function returns an expression sequence consisting of the square of all operands of an arbitrary expression

```
> sqall:=x->seq(z^2,z=x):
> sqall([a,b,c,d,e]);
```

$$a^2, b^2, c^2, d^2, e^2$$

```
> sqall(a^2+b*c);
```

$$a^4, b^2c^2$$

An expression sequence can be used as the second argument of seq, provided it is enclosed in parentheses

```
> seq(z-1,z=(1,2*4,joe));
```

$$0, 7, joe - 1$$

The parentheses are not needed if the expression sequence is represented symbolically by a variable (including a ditto variable)

```
> es:=1,8,joe:
> seq(z-1,z=es);
```

$$0, 7, joe - 1$$

In chapter 3, we have introduced the construct `map(f,A)` to obtain the image of a set A under a function `f`. The standard library function `map` allows for more general usage. Its second argument can be any expression, not just a set: `map` will then act on its top-level operands, while preserving the data type. It is instructive to compare the behaviour of the functions `map` and `seq` when called with similar arguments

```
> f:=x->x^2:
> map(f,a+b+c);
```

$$a^2 + b^2 + c^2$$

```
> seq(f(s),s=a+b+c);
```

$$a^2, b^2, c^2$$

While `map` preserves the structure of the expression, `seq` transforms it into an expression sequence.

If the first operand `f` of `map` requires more than one argument, then the additional arguments $x_2, x_3, \ldots$ must be supplied after the target expression: `map(f,expr,x2,x3,..)`. For illustration, we consider a list L containing numerical data. We wish to replace any element of L which is larger than the last element by the latter. This is accomplished by the following function

```
> cap:=L->map(min,L,op(nops(L),L));
> cap([1,2,3,4,5,6,7,5]);
```

$$[1, 2, 3, 4, 5, 5, 5, 5]$$

The last element of the list L is given by `op(nops(L),L)`. The first argument of the function `min` runs through all the elements of L, while the second argument is supplied as the first optional argument of `map`.

Another useful utility function is `select`, which selects the operands of an expression, according to some criteria. The latter is specified by a Boolean function, which is applied to all operands of the target expression. The operands giving the value *true* are those to be selected. The target expression is usually a list or a set.

To compute the primes between a and b, we first construct the list of integers between a and b, which is `[$a..b]`, and then select the primes in the list with the Boolean function `isprime`.

```
> select(isprime,[$100..110]);
```

$$[101, 103, 107, 109]$$

The output is a list, the same data type of the second argument of `select`.

The syntax of `select` resembles that of `map`. Its basic form is `select(f, expr)`, where `f` is the symbolic name of a Boolean function (or an explicit function definition using the arrow operator), while `expr` is an arbitrary expression. The function `select` returns an expression of the same data type as `expr`, containing only the selected operands. If the process of selection is such as to change the data type of `expr`, an error termination will occur. As for `map`, if `f` requires more than one argument, then the additional arguments $x_2, x_3, \ldots$ must be supplied after the target expression: `select(f,expr,x2,x3,..)`. The following function selects the non-negative operands of an arbitrary expression

```
> NonNeg:=A->select(x->evalb(x>=0),A):
> NonNeg({-3,5,0,-33});
```

$$\{0, 5\}$$

The function `remove`, which is complementary to `select`, removes (rather than selects) the operands which match given criteria. Its syntax is identical to that of `select`. It is plain that everything that can be done with `remove` can also be done with `select`, and vice versa. For instance, if L is a list of integers, the expressions

```
> remove(isprime,L);
> select(x->not isprime(x), L);
```

will have the same value.

The function `evalf` also acts on operands, but it does so more selectively than `map`, `seq`, or `select`.

```
> evalf(33*x^(1/3)+22*x/4-13/5);
```

$$33. \, x^{1/3} + 5.500000000 \, x - 2.600000000$$

Note that `evalf` has converted into `float` only the coefficients of the above algebraic expression, leaving the `exponent` unevaluated. The function `evalf` will act in a similar fashion on the real and imaginary part of a complex number.

Example 6.1. Let $L = [[x_1, y_1], \ldots, [x_n, y_n]]$ be a list of points in the Cartesian plane. The following function returns the list $[x_1, \ldots, x_n]$ of the corresponding x-coordinates

```
> xcoord:=L->map(p->op(1,p),L):
```

Example 6.2. We generate the set of the irreducible factors of the polynomial $x^{12} - 1$

```
> {seq(f,f=factor(x^12-1))};
```

$$\{x - 1, x^2 + 1, x + 1, x^4 - x^2 + 1, x^2 - x + 1, x^2 + x + 1\}$$

The above method would not work for the polynomial $x^4 + 1$. Why?

Example 6.3. Let x be a real number, and let A be a set (list) of real numbers. We construct a function of x and A whose value is the subset (sublist) of A constituted by all the elements of A that are smaller than x.

```
> L:=(x,A)->select((y,x)->evalb(y < x),A,x):
> A:={3,44,-2,7,23,0,-34}:
> L(7,A);
```

$$\{0, -2, 3, -34\}$$

If A contains symbolic data (i.e., π or $\sqrt{2}$), then the Boolean expression performing the selection will have to be modified to `evalb(evalf(y < x))`, in order to be evaluated. The following variant of the above construct uses `nops` to count the elements of an arbitrary expression A that are equal to x.

```
> N:=(x,A)->nops(select((y,x)->evalb(y=x),A,x):
```

Example 6.4.   We construct a function which returns the number of twin prime pairs $(p, p + 2)$ where p lies between a and b and $a \le b$ (see (3.6)). The relevant Boolean function selects those integers x such that x and $x+2$ are both primes. The counting is done by `nops`.

```
> ntwins:=(a,b)->
>         nops(select(x->isprime(x) and isprime(x+2), [$a..b])):
> ntwins(100,200);
```

$$7$$

Example 6.5.   The following is a homemade version of the intrinsic function `intersect`, computing the intersection between two sets.

```
> MyIntersect:=(A,B)->select(member,A,B):
```

Note that `member(x,S)` is a Boolean function of two arguments, an expression x and a set S. In the above construct x runs through all operands of A, while S is supplied as the optional third argument of `select`. This function is slightly more flexible than `intersect`: explore its behaviour.

Visible points

We provide a geometrical interpretation of the concept of relative primality, introduced in section 2.4. A point on the plane whose coordinates are integers will be called an *integer point.* We denote by $\mathbb{Z}^2$ the collection of all integer points. An integer point (x, y) is *visible from the origin* if the segment connecting (x, y) to $(0, 0)$ does not contain any integer point beside the endpoints (figure 6.5).

The following result relates the geometrical and the arithmetical characterizations of visible points.

Theorem 6 *A point (x, y) of $\mathbb{Z}^2$ is visible from the origin if and only if x and y are relatively prime.*

If x and y have a common divisor $d > 1$, then $x = dx'$ and $y = dy'$ for some integers x' and y', and since $|x'| < |x|$ and $|y'| < |y|$, the point (x', y')

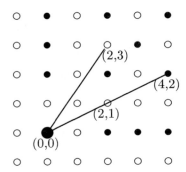

Figure 6.5: Visible and invisible points from the origin $(0,0)$ of $\mathbb{Z}^2$, represented by circles and solid circles, respectively. The point $(2,3)$ is visible, while the point $(4,2)$ is not, since its view is obstructed by $(2,1)$.

obstructs the view of (x,y) from the origin (think about it). This proves the 'if' bit of the theorem.

To prove the 'only if' implication, we assume that x and y are coprime. If $|x|$ or $|y|$ is equal to 1, then (x,y) is clearly visible (draw a picture!), so we shall assume that $|x|, |y| > 1$. We proceed by contradiction, and assume that (x,y) is not visible. Then there exists a point (x',y') lying between (x,y) and the origin, which implies that $xy' = yx'$. If x and x' were coprime, then from the fundamental theorem of arithmetic, we would conclude that x divides y (why?), contrary to the assumption that x and y are coprime. So the greatest common divisor d of x and x' is greater than 1, and we write $x = ad$, $x' = bd$. Now, a and b are coprime, and moreover $a > 1$, since $1 \leq |x'| < |x|$. So a is also a proper divisor of y, and therefore x and y are not coprime, a contradiction.

We wish to construct the (in)visible points contained within the square region $0 \leq x, y, \leq n$, where n is an integer. We shall represent an integer point as a two-element Maple list [x,y]. We begin by constructing the Boolean characteristic function of the visible points

```
> chi:=z->evalb(igcd(op(1,z),op(2,z))=1):
```
Next we construct the set of all integer points in a suitably large square region

```
> n:=50:
```
```
> IntPts:={seq(seq([x,y],y=0..n),x=0..n)}:
```
The visible points are extracted from the above set using select

```
> VisPts:=select(chi,IntPts):
```
Note that VisPts is also a set, because select retains the data type of the

target expression. The invisible points are the difference between the above
two sets

```
> InvisPts:=IntPts minus VisPts:
```
To plot (in)visible points, we must first convert the corresponding data set
into a list.

```
> plot([op(VisPts)],style=POINT,scaling='constrained');
```

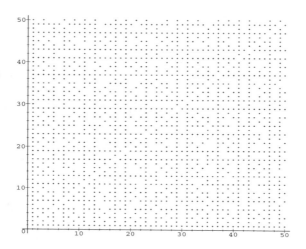

Considering horizontal and vertical lines, we see that the density of visible
points is maximal whenever the corresponding coordinate is a prime number, and minimal when it is an integer with a large number of divisors, e.g.,
30. (A more accurate formulation of this phenomenon requires knowledge
of *Euler's ϕ-function*.)

To count visible points, we use `nops`

```
> nops(VisPts);
```

$$1549$$

Since our region contains $51^2 = 2601$ integer points, we see that approximately 60% of them are visible from the origin. We shall return to the
question of the density of visible points in section 8.4, after having developed a more effective method for counting such points.

Exercises

Exercise 6.4. Each expression E on the left is transformed into the expression on the right by `map(f,E)`, for a suitable function `f`. Construct such user-defined function `f` in each case.

	E	$\texttt{map(f, E)}$
(a)	$[0, 1, 2, 3]$	$[100, 101, 102, 103]$
(b)	$a + b + c + d$	$1/a + 1/b + 1/c + 1/d$
(c)	$\{-20, 10, -10, -30, 20\}$	$\{10, 20, 30\}$
(d)	$[0, 1, 1, 1, 0, 1, 0, 1]$	$[1, 0, 0, 0, 1, 0, 1, 0]$
(e)	$[a + b, c + d, e + f]$	$[a, b, c, d, e, f]$
(f)	$[a + b, c + d, e + f]$	$[ab, cd, ef]$
(g)	$[a^3, b^4, a^7]$	$[a/(3 + a),\ b/(4 + b),\ a/(7 + a)]$
(h)	$[[a], [[b]], [[[c]]], [[[[d]]]]]$	$[a,\ b,\ [c],\ [[d]]]$
(i)	$[[a], [d, e, f], [g, h, i, j], [b, c]]$	$[1, 3, 4, 2]$
(j)	$[[a], [b, c], [d, e, f, g], [h, i, j]]$	$[a, c, g, j]$

Exercise 6.5. The following function generates the list of the prime divisors of an integer $n > 1$.

```
> PrimeDiv:=n->[seq(expand(op(1,x)),x=ifactor(n))]:
> PrimeDiv(18800);
```

$$[2, 5, 47]$$

However, this function does not produce the desired answer for all values of n. Characterize the set of integers n for which it does.

Exercise 6.6. A Maple list L contains numerical data, representing the elements of a sequence.

(a) Plot the elements of L.

(b) Determine the number of negative elements of L.

(c) Let $a_1, \ldots, a_k$ be the *distinct* elements of L, ordered arbitrarily. For every $i = 1, 2, \ldots, k$, let n_i be the frequency of occurrence of a_i, i.e., the number of times a_i appears in L. Construct a function `freq(L)` which returns the frequency of each distinct element of L, with the following format

$$[[a_1, n_1], [a_2, n_2], \ldots, [a_k, n_k]].$$

Chapter 7

Polynomials and rational functions

Polynomials and rational functions are Maple's favourite data types. In this chapter we look at their arithmetical properties, which resemble those of integers and rationals, respectively.

7.1 Polynomials

A polynomial is an object such as $x^3 - 3x - 1$, consisting of integers combined with copies of an indeterminate x, raised to various integral powers. More formally, let $a_0, a_1, \ldots, a_{n-1}, a_n$ be $n+1$ elements of a set K, equipped with sum, subtraction, and multiplication (K is a *ring* or a *field*), which we will assume to be one of $\mathbb{Z}, \mathbb{Q}, \mathbb{R}$, or $\mathbb{C}$. A *polynomial over K* in the *indeterminate x* with *coefficients $a_0, \ldots, a_n$* is an expression

$$p = a_0 x^0 + a_1 x^1 + a_2 x^2 + \cdots + a_{n-1} x^{n-1} + a_n x^n \qquad a_k \in K.$$

The set of all polynomials over K in the indeterminate x will be denoted by $K[x]$. The integer n is the *degree* of p (denoted by $\deg(p)$), and a_n is the *leading coefficient*. The term x^0 associated with the *constant coefficient a_0* is usually omitted, and so is the exponent 1 in x^1: $\ p = a_0 + a_1 x + a_2 x^2 + \cdots$.

Example 7.1. The polynomial $p = 3 - x^2 + 2x^3$ is an element of $\mathbb{Z}[x]$. Its coefficients are

$$a_0 = 3, \quad a_1 = 0, \quad a_2 = -1, \quad a_3 = 2.$$

The degree of p is 3, and its leading coefficient is 2. Maple knows this
> p:=3-x^2+2*x^3;

$$p := 3 - x^2 + 2x^3$$

127

```
> degree(%),lcoeff(%);
```

$$3, 2$$

Note that for the notion of degree to be well-defined, we have tacitly required that a_n be nonzero, if n is positive. (It would not make much sense to say, for instance, that the polynomial $-1 + 3x + 0x^{29}$ is a polynomial of degree 29, with leading coefficient zero!)

The polynomials of degree 0 in $K[x]$, given by $p = a_0x^0$, can be naturally associated with the element a_0 of K (even though a_0 and a_0x^0 are not quite the same thing, as we shall see). Then an element of K can be thought of as an element of $K[x]$, and we have the inclusion $K \subset K[x]$. Moreover, if $K \subset L$, then $K[x] \subset L[x]$, because every polynomial with coefficients in K is also a polynomial with coefficients in L. So, for instance, the chain of inclusions (5.12) implies the associated chain of inclusions for polynomials

$$\mathbb{Z}[x] \subset \mathbb{Q}[x] \subset \mathbb{R}[x] \subset \mathbb{C}[x].$$

Example 7.2.   Examples of polynomials.

p	$\deg(p)$	a_0	a_1	a_2	a_3	a_4	$K[x]$
x^4	4	0	0	0	0	1	$\mathbb{Z}[x]$
$3 - x^2 + 2x^3$	3	3	0	-1	2		$\mathbb{Z}[x]$
$-21/5$	0	$-21/5$					$\mathbb{Q}[x]$
$(x^2 + 6)/3$	2	2	0	$1/3$			$\mathbb{Q}[x]$
$\sqrt{5}x^2 - x$	2	0	-1	$\sqrt{5}$			$\mathbb{R}[x]$
$x^3 - ix/2$	3	0	$-i/2$	0	1		$\mathbb{C}[x]$

7.2   Polynomial arithmetic

A polynomial is in the first instance an arithmetical object, not a function of the indeterminate. We are not interested in substituting any specific value for x. (Accordingly, we tend to define a polynomial as a Maple expression, rather than a Maple function.) The role of the indeterminate is merely that of organizing the sequence of coefficients.

Arithmetically, polynomials behave much like integers. The elements of $K[x]$ can be summed, subtracted, and multiplied unrestrictedly, and this results from the fact that the elements of K, as well as the indeterminate x, enjoy that same property.

Addition and subtraction

The coefficient of the term x^k of the sum (difference) of two polynomials is given by the sum (difference) of the corresponding coefficients of the summands.

We begin with an example. Let $p = -5 + x$ and $q = -x + x^2$. We write all coefficients explicitly, matching degrees

$$p = -5 + 1\,x + 0\,x^2 \qquad q = 0 - 1\,x + 1\,x^2.$$

Then

$$
\begin{aligned}
p + q &= (-5 + 0) + (1 - 1)x + (0 + 1)x^2 = -5 + x^2 \\
p - q &= (-5 - 0) + (1 - (-1))x + (0 - 1)x^2 = -5 + 2x - x^2.
\end{aligned}
$$

```
> p:=-5+x:
> q:=-x+x^2:
> p+q,p-q;
```

$$-5 + x^2,\ -5 + 2x - x^2$$

As for integers, the sum and subtraction are worked out automatically by Maple. Note that sum and subtraction may lower the degree, by cancelling out the leading coefficient

```
> p:=2+2*x^5:
> q:=x^3-2*x^5:
> degree(p),degree(q),degree(p+q),degree(p-q);
```

$$5,\ 5,\ 3,\ 5$$

In this example, a polynomial of degree 3 resulted from the sum of two polynomials of degree 5. It should be clear that sum or subtraction cannot increase the degree; so the rule relating the degree of the summands to that of the sum is given by

$$\deg(p \pm q) \leq \max(\deg(p),\ \deg(q)).$$

To define the operations of sum and subtraction in the general case, it is convenient to regard a polynomial as an *infinite* sum

$$p = \sum_{i=0}^{\infty} a_i\, x^i \tag{7.1}$$

where only *finitely many* coefficients a_i are allowed to be nonzero. Then the sum (7.1) only 'looks' infinite, but is actually finite. With this device the sum/subtraction of two polynomials can be written without being concerned about their degrees

$$p \pm q = \sum_{i=0}^{\infty} a_i\, x^i \pm \sum_{i=0}^{\infty} b_i\, x^i = \sum_{i=0}^{\infty} (a_i \pm b_i)\, x^i.$$

We see that the coefficients of the sum (difference) of p and q are the sum (difference) of the respective coefficients. Is $p \pm q$ as defined above a polynomial? Yes, because the rightmost sum also has finitely many nonzero coefficients, as easily verified (do it!).

Multiplication

We begin with an example. Let $p = 1 - 2x + x^2$ and $q = 3 + 5x^2 - x^3$. Then, writing all coefficients explicitly, we have

$$
\begin{aligned}
p \cdot q &= (1 - 2x + 1x^2) \cdot (3 + 0x + 5x^2 - 1x^3) \\
&= (1 \cdot 3) + (-2 \cdot 3 + 1 \cdot 0)x + (1 \cdot 3 + (-2) \cdot 0 + 1 \cdot 5)x^2 \\
&\quad + (1 \cdot (-1) + (-2) \cdot 5 + 1 \cdot 0)x^3 \\
&\quad + ((-2) \cdot (-1) + 1 \cdot 5)x^4 + (1 \cdot (-1))x^5 \\
&= 3 - 6x + 8x^2 - 11x^3 + 7x^4 - x^5.
\end{aligned}
$$

To find the coefficient of x^k in the product of two polynomials, we multiply the coefficient of x^j in one polynomial with the coefficient of x^{k-j} in the other, and then add all these products, for $j = 0, 1, \ldots, k$.

```
> p:=1-2*x+x^2:
> q:=3+5*x^2-x^3:
> p*q;
```

$$
(1 - 2x + x^2)(3 + 5x^2 - x^3)
$$

```
> expand(%);
```

$$
3 + 8x^2 - 11x^3 - 6x + 7x^4 - x^5
$$

Contrary to what happens for numbers, Maple does not expand the product of polynomials automatically. There is a good reason for this, as we shall see. To force expansion, one must use the function **expand**, which multiplies out the factors, and then carries out standard simplifications. (Note that Maple does not necessarily represent the polynomial in ascending powers of the indeterminate.) The only instance in which Maple expands a product automatically is when one of the operands has degree zero.

```
> 9*(x+2);
```

$$
9x + 18
$$

The formula for polynomial multiplication is a bit more complicated than that for addition. Using again the representation of polynomials as infinite sums with finitely many nonzero coefficients, we let

$$
p \times q = \sum_{i=0}^{\infty} a_i x^i \times \sum_{i=0}^{\infty} b_i x^i = \sum_{i=0}^{\infty} c_i x^i
$$

where (see exercises)

$$
c_i = \sum_{j=0}^{i} a_j b_{i-j} = \sum_{j+k=i} a_j b_k, \qquad i = 0, 1, \ldots. \tag{7.2}
$$

The degree of the product of two polynomials is equal to the sum of the degrees of the operands

$$\deg(p \cdot q) = \deg(p) + \deg(q).$$

Division

Like for integers, the ratio of two polynomials may or may not be a polynomial. In $K[x]$ — which is structurally identical to $\mathbb{Z}$ — division cannot always be carried out, and one needs to introduce the corresponding notion of divisibility.

Given two polynomials p and q in $K[x]$, we say that p divides q if there exists a polynomial h in $K[x]$ such that $ph = q$. As for integers, we write $p|q$ to denote divisibility.

For instance, let $p = 1 - x$ and $q = -1 + x^2$. Then p divides q in $\mathbb{Z}[x]$, because there exists the polynomial $h = -1 - x$ such that $ph = q$.

```
> p:=1-x:
> q:=-1+x^2:
> q/p;
```

$$\frac{-1 + x^2}{1 - x}$$

```
> simplify(%);
```

$$-x - 1$$

Again, Maple refuses to carry out the simplification automatically, and we have to use the function simplify for this purpose.

A greatest common divisor of two polynomials is a polynomial of highest degree dividing both. A least common multiple of two polynomials is a polynomial of smallest degree divisible by both. They are unique up to a constant factor.

Example 7.3. *The divisors of a polynomial come in pairs.* Indeed, if p divides q, then also q/p divides q, and vice versa. The argument is identical to that given in chapter 2 for the integers. Thus, in the above example, both $p = 1 - x$ and $q/p = -1 - x$ divide q. As for the integers, the number of divisors of a polynomial p is even, except when p is a square of another polynomial.

The analogy between polynomials and integers can be carried out further.

Theorem 7 *If f and g are polynomials over $\mathbb{Q}$, $\mathbb{R}$, or $\mathbb{C}$, and g is nonzero, then one can find unique polynomials q and r such that*

$$f = g\,q + r \tag{7.3}$$

where r has smaller degree than g.

The polynomials q and r are the *quotient* and the *remainder*, respectively, on dividing f by g.   (Compare with the analogous construction given in equation (2.10).) Note that the above theorem does not apply to polynomials in $\mathbb{Z}[x]$. To see why, let us consider the case $f(x) = x^2 - 1$ and $g(x) = 2x - 2$. Then g divides f in $\mathbb{Q}[x]$, but not in $\mathbb{Z}[x]$

$$x^2 - 1 = (2x - 2)\left(\frac{1}{2}x + \frac{1}{2}\right).$$

In $\mathbb{Z}[x]$, the equation (7.3) gives $q(x) = 0$ and $r(x) = g(x)$, with the degree of $r(x)$ equal to that of $g(x)$. It turns out that Theorem 7 can be extended to $\mathbb{Z}[x]$, provided that we consider only *monic polynomials*, that is, polynomials with leading coefficient equal to 1.

Quotient and remainder may be computed using the long division algorithm from elementary algebra. The working for the case $f = x^5 - 1$ and $g = x^2 - x - 1$ is displayed below

$$
\begin{array}{l}
 x^5 -1 \;\Big|\; \underline{x^2 - x - 1} \\[2pt]
\underline{x^5 - x^4 - x^3} \\
 x^4 + x^3 x^3 + x^2 + 2x + 3 \\
 \underline{x^4 - x^3 - x^2} \\
 2x^3 + x^2 \\
 \underline{2x^3 - 2x^2 - 2x} \\
 3x^2 - 2x - 1 \\
 \underline{3x^2 - 3x - 3} \\
 5x + 2
\end{array}
$$

One finds that $q = x^3 + x^2 + 2x + 3$ and $r = 5x + 2$. Equation (7.3) becomes

$$x^5 - 1 = (x^2 - x - 1)(x^3 + x^2 + 2x + 3) + 5x + 2.$$

What are the 'prime' polynomials? A polynomial over $K[x]$ is said to be *irreducible* if it cannot be expressed as the product of two polynomials of $K[x]$ of smaller degree. Thus, the polynomial $p = x^2 - x - 2$ of degree 2 is not irreducible in $\mathbb{Q}[x]$, because $p = (x - 2)(x + 1)$ is the product of two polynomials in $\mathbb{Q}[x]$ of degree 1. For this definition to work, we have to agree that the polynomials of degree 0 are not considered irreducible, much in the same way as the integer 1 is not considered prime.

Theorem 8 (the fundamental theorem of polynomial arithmetic). *Every polynomial over* $\mathbb{Q}$, $\mathbb{R}$, *or* $\mathbb{C}$ *can be expressed as a product of irreducibles, and this factorization is unique up to constant factors and the order of the factors.*

The formulation of this theorem is virtually identical to that of the fundamental theorem of arithmetic. The only extra condition is that constant factors are to be ignored when considering the question of uniqueness of the decomposition into irreducibles, in accordance with our definition of irreducibility. To illustrate this point, let us consider the polynomial $2 + 2x = 2(1 + x)$. If we do not ignore the constant factor 2, which is a polynomial of degree zero, then we would conclude that this polynomial has at least two distinct representations into irreducibles.

Maple has functions for divisibility of polynomials over $\mathbb{Q}$ (and $\mathbb{Z}$) which mimic closely the corresponding functions for integers.

DIVISIBILITY FUNCTIONS

FOR INTEGERS AND POLYNOMIALS OVER $\mathbb{Q}$

integers	polynomials
iquo(a,b)	quo(p,q,x)
irem(a,b)	rem(p,q,x)
igcd(a,b)	gcd(p,q)
ilcm(a,b)	lcm(p,q)
ifactor(a)	factor(p)

Example 7.4. Consider the following factorizations
```
> ifactor(6),factor(1-x^3);
```
$$(2)(3), \ -(x - 1)(x^2 + x + 1)$$

In the same way that 2 and 3 are the prime factors of 6 in $\mathbb{Z}$, so $1 - x$ and $1 + x + x^2$ are the irreducible factors of $1 - x^3$ in $\mathbb{Q}(x)$.
```
> ifactor(6),factor(6);
```
$$(2)(3), \ 6$$

Maple factors the integer 6 in $\mathbb{Z}$, but it does not factor the degree zero polynomial $6 = 6\,x^0$ in $\mathbb{Q}[x]$, conforming with the stipulation that such polynomials are excluded from the factorization process (cf. remark before Theorem 8). This example shows the extent to which the integer n and the polynomial $n = n\,x^0$ of degree 0 cannot be identified.

Example 7.5. *Quotient and remainder.*
```
> f:=x^4+x+1:
> g:=x^2-1:
> quo(f,g,x);
```
$$x^2 + 1$$
```
> rem(f,g,x);
```
$$x + 2$$

```
> evalb(expand((g*%%+%)-f)=0);
```

$$true$$

The last statement verifies the equality (7.3).

Example 7.6. *Divisibility and remainder.* As we did for the integers, we can use **rem** to test divisibility of polynomials. The following function returns *true* if q divides p in $\mathbb{Q}[x]$, and *false* otherwise.

```
> dvd:=(p,q,x)->evalb(rem(p,q,x)=0):
```

Sequences of polynomials

A sequence $g_0, g_1, g_2, \ldots$ of polynomials in $K[x]$ is a function that associates to every natural number t an element of $K[x]$

$$g : \mathbb{N} \to K[x] \qquad g := t \mapsto g_t.$$

Note that in this context g is a function of t, not a function of x. For instance, if we let

$$g : \mathbb{N} \to \mathbb{Z}[x] \qquad g := t \mapsto x^2 - tx + t - 1$$

we obtain a sequence of polynomials over $\mathbb{Z}$, all of degree 2

$$x^2 - 1,\ x^2 - x,\ x^2 - 2x + 1, \ldots$$

```
> g:=t->x^2-t*x+t-1:
> seq(g(t),t=0..5);
```

$$x^2 - 1,\ x^2 - x,\ x^2 - 2\,x + 1,\ x^2 - 3\,x + 2,\ x^2 - 4\,x + 3,\ x^2 - 5\,x + 4$$

In the above example the variable x is not passed to the function g as an argument. We say that x is a *global variable,* meaning that its value can be modified from outside the function.

```
> x:=-1:
> seq(g(t),t=0..5);
```

$$0,\ 2,\ 4,\ 6,\ 8,\ 10$$

Global variables can lead to unpredictable results and should be used sparingly.

Sequences of polynomials can be defined *recursively,* in a manner analogous to sequences of numbers. In the simplest instances, a recursive sequence is defined by the initial condition g_0, which is a given polynomial $f \in K[x]$, and by a rule that defines the element g_{t+1} of the sequence as a given function of g_t. Such function is arbitrary, and in particular it may involve other polynomials in $K[x]$.

$$g_0 = f \qquad g_{t+1} = F(g_t), \qquad t \geq 0.$$

For instance, let $K = \mathbb{Z}$, and let

$$g_0 = 1 + x \qquad g_{t+1} = x^2 \cdot g_t^2 - 1, \qquad t \geq 0.$$

Then

$$
\begin{aligned}
g_1 &= x^2 \, g_0^2 - 1 = x^2(x+1)^2 - 1 = x^4 + 2x^3 + x - 1, \\
g_2 &= x^2 \, g_1^2 - 1 = x^2(x^2(x+1)^2 - 1)^2 - 1 \\
&= x^{10} + 4\,x^9 + 6\,x^8 - x^6 + 4\,x^7 - 4\,x^5 - 2\,x^4 + x^2 - 1, \\
&\;\;\vdots
\end{aligned}
$$

Example 7.7. *Polynomials as sequences of coefficients.* For a polynomial, the role of the indeterminate is simply that of organizing a finite sequence of coefficients. We can dispose of the indeterminate altogether and represent polynomials as sequences of elements of the underlying set K. Thus, the polynomial $p = 3 - x^2 + 2x^3$ in $\mathbb{Z}[x]$ can be represented as the finite sequence of integers $p = (a_0, a_1, a_2, a_3) = (3, 0, -1, 2)$. Better still, we could regard p as an *infinite* sequence of elements of $\mathbb{Z}$, with only a *finite* number of them being different from zero: $p = (3, 0, -1, 2, 0, 0, \ldots)$. With this device, the arithmetic of polynomials is described in very elegant terms. For instance, if $p = (a_0, a_1, \ldots)$ and $q = (b_0, b_1, \ldots)$ are polynomials over K, then $p + q = (a_0 + b_0, a_1 + b_1, \ldots)$. Furthermore, the elements of K are identified with the sequences where only the first term is nonzero

$$p \in K \iff p = (a, 0, 0, 0, \ldots).$$

Polynomials in several indeterminates *

We have considered polynomials whose coefficients are numbers. However, all that is required from the coefficients is the possibility of performing the arithmetical operations of sum, subtraction, and multiplication. But these properties are also enjoyed by polynomials, so it makes sense to consider polynomials whose coefficients are themselves polynomials.

For illustration, let us consider the expression

$$p = (c^3 + 2\,c^2 + c - 1) + x\,(c^2 + 2\,c + 1) + x^3\,(3\,c - 1) + x^4.$$

This is a polynomial of degree 4 in the indeterminate x: $p = a_0 + a_1\,x + a_2\,x^2 + a_3\,x^3 + a_4\,x^4$. Its coefficients a_k are polynomials in the indeterminate c

$$a_0 = c^3 + 2\,c^2 + c - 1, \quad a_1 = c^2 + 2\,c + 1, \quad a_2 = 0, \quad a_3 = 3\,c - 1, \quad a_4 = 1.$$

Thus, the coefficients of p belong to $\mathbb{Z}[c]$, so that p belongs to $\mathbb{Z}[c][x]$. It is customary to denote the latter set by $\mathbb{Z}[c, x]$, which underlines the fact that the indeterminates x and c have equal status. This means that we could equally regard this polynomial as a polynomial in the indeterminate c

$$ p(c) = (-1 + x - x^3 + x^4) + c\,(1 + 2\,x + 3\,x^3) + c^2\,(x + 2) + c^3. $$

This time $p(c) = b_0 + b_1\,c + b_2\,c^2 + b_3\,c^3$ is a polynomial of degree 3, with coefficients in $\mathbb{Z}[x]$

$$ b_0 = -1 + x - x^3 + x^4, \quad b_1 = 1 + 2\,x + 3\,x^3, \quad b_2 = x + 2, \quad b_3 = 1. $$

(This example should make it clear that $\mathbb{Z}[x, c] = \mathbb{Z}[c, x]$; think about it.) The Maple intrinsic function **degree** accepts an optional second argument specifying the indeterminate with respect to which the degree is to be evaluated

```
> p:=(c^3+2*c^2+c-1) + x*(c^2+2*c+1)+ x^3*(3*c-1)+x^4:
> degree(p,x),degree(p,c);
```

$$ 4,\ 3 $$

Exercises

Exercise 7.1. Prove the formula (7.2) for the coefficients of the product of two polynomials.

Exercise 7.2. Prove that the degree of the product of two polynomials is equal to the sum of the degrees of the operands.

Exercise 7.3. Show that $x^4 + x^3 + x^2 + x + 1$ and $x^2 + x + 1$ are relatively prime.

Exercise 7.4. List *all* common divisors of $x^6 - x^5 - 2x^4 + x^3 + x^2 + 2x - 2$ and $x^5 - x^3 - x^2 + 1$ in $\mathbb{Z}[x]$.

Exercise 7.5. Let

$$ f = x^6 - x - 1 \qquad g = x^3 + x + 1. $$

Let q and r be the quotient and remainder, respectively, of the division of f by g. Verify that

$$ \frac{f}{g} = q + \frac{r}{g}. $$

Exercise 7.6. Consider the *recursive* sequence of polynomials in $\mathbb{Z}[x]$

$$ g_0 = x \qquad g_{t+1} = 1 + x \cdot g_t, \qquad t \geq 0. $$

Determine the smallest value of t for which g_t is reducible.

Exercise 7.7. Consider the sequence of polynomials over $\mathbb{Z}$

$$p_n(x) = x^{2\,n-1} + 1 \qquad n \geq 1.$$

Let $q_n(x)$ be the quotient of the division of $p_n(x)$ by $x + 1$. Determine the sequence $q_n(x)$, $n \geq 1$.

Exercise 7.8. Consider the sequence of polynomials over $\mathbb{Z}$

$$c_n(x) = x^n - 1 \qquad n \geq 1.$$

The purpose of this exercise is to study how the elements of this sequence factor into the product of irreducible polynomials.

(a) Construct a Maple function `c(n)`, whose value is $c_n(x)$, *in factored form*. Hence factor $c_n(x)$ for $n = 6, 9, 12, 13$. In each case, count the number of irreducible factors, and then verify that such number is also the value of the Maple function `nops` applied to the factored expression of $c_n(x)$. (This does not work when $n = 1$, though — why?)

(b) Let n be a positive integer, and let I_n be the number of irreducible factors of $c_n(x)$. So we have the sequence of positive integers

$$I_1, I_2, I_3, \ldots \qquad n \geq 1.$$

Using `nops`, construct a function called `nf(n)` whose value is I_n for $n \geq 2$ (this function will give the wrong answer for $n = 1$). Test it. Hence compute I_{83} and I_{84}.

(c) Plot the first 30 elements of the sequence $n \mapsto I_n$, including the correct entry for $n = 1$. Observe the very irregular behaviour of this sequence.

(d) Let I_n be as above. There exists a close relationship between I_n and the divisors of n. Discover what it is, and then express your findings in the form of a conjecture.

Exercise 7.9*. Let $f(x)$ be a polynomial. A number α is a *root* of $f(x)$ if $f(\alpha) = 0$. Thus, $\sqrt{2}$ is a root of $f(x) = x^2 - 2$, and so is $-\sqrt{2}$.

For any positive integer n, we let

$$f_n(x) = x^n - 1 \qquad \alpha_{n,k} = e^{2\pi i k/n}, \qquad k = 0, 1, \ldots, n - 1.$$

Then $\alpha_{n,k}$ is a root of $f_n(x)$ for any choice of k, because

$$f_n(\alpha_{n,k}) = \alpha_{n,k}^n - 1 = e^{2\pi i k} - 1 = 1 - 1 = 0.$$

Now let $n = 15$. The polynomial $f_{15}(x)$ factors into the product of irreducible polynomials, the largest of which has degree 8. Determine the 8 values of k corresponding to the roots of the latter. (You must do so without performing floating-point calculations.)

7.3   Rational functions

The lack of closure of the integers under division led to the introduction of the rational numbers. With polynomials there is a similar need of enlarging the set $K[x]$ in order to achieve closure with respect to division. Thus, the ratio $(x^2 - 1)/(x - 1) = x + 1$ is a polynomial, because $x - 1$ divides $x^2 - 1$

```
> (x^2-1)/(x-1):%=simplify(%);
```

$$\frac{x^2 - 1}{x - 1} = x + 1$$

However, the ratio $(x^2 - 1)/(x - 2)$ is not a polynomial, since $x - 2$ does not divide $x^2 - 1$.

```
> simplify((x^2-1)/(x-2));
```

$$\frac{x^2 - 1}{x - 2}$$

In this case we speak of a *rational function.*

The rational functions over K in the indeterminate x — denoted by $K(x)$ — is the set of fractions whose numerator and denominator are elements of $K[x]$, and the denominator is nonzero. This construction is analogous to the construction of $\mathbb{Q}$ from $\mathbb{Z}$. In particular, we have $K[x] \subset K(x)$, since every polynomial over K is also a rational function with denominator $1 = 1 \cdot x^0$, and the latter is itself a polynomial over K. (This is due to the fact that every *field* K contains 1; see section 12.3.) So we have the chain of inclusions

$$\mathbb{Q}(x) \subset \mathbb{R}(x) \subset \mathbb{C}(x).$$

Note that we have not mentioned the set $\mathbb{Z}(x)$, since the latter is equal to $\mathbb{Q}(x)$. (Why?)

Recall that Maple simplifies rational numbers in such a way that the denominator is positive, and the numerator and denominator are relatively prime. Maple does not simplify rational functions, unless the simplification is obvious, that is, unless the quantity to be simplified is already factored. The intrinsic function `simplify` represents a rational function over $\mathbb{Q}$ so that numerator and denominator are relatively prime and their coefficients are integer (so does the function `normal`).

```
> p:=(x-1)*(x-2)*(x-3):
> q:=(x-1)*(x+1):
> p/q;
```

$$\frac{(x - 2)(x - 3)}{x + 1}$$

```
> p/expand(q);
```

$$\frac{(x - 1)(x - 2)(x - 3)}{x^2 - 1}$$

```
> simplify(%);
```

$$\frac{(x-2)(x-3)}{x+1}$$

Simplification is not always desirable. For instance, the rational function

$$\frac{x^{1000000} - 1}{x - 1} = x^{999999} + x^{999998} + \cdots + x + 1$$

which is defined by two polynomials with four nonzero coefficients in all, 'simplifies' to a polynomial with one million nonzero coefficients!

7.4   Basic manipulations

We have already considered the functions `expand` and `factor`
```
> p:=(x-1)*(x^2+2*x+1):
> expand(p);
```

$$x^3 + x^2 - x - 1$$

```
> factor(%);
```

$$(x - 1)(x + 1)^2$$

```
> op(%);
```

$$x - 1, (x + 1)^2$$

With `expand`, all products are distributed over sums, and like terms are collected

$$\begin{aligned}
(x + 2)(x + 3) \quad &\longrightarrow x(x + 3) + 2(x + 3) \\
&\longrightarrow x^2 + 3x + 2x + 6 \\
&\longrightarrow x^2 + 5x + 6
\end{aligned}$$

or

$$\begin{aligned}
(x + 2)(x + 3) \quad &\longrightarrow x(x + 2) + 3(x + 2) \\
&\longrightarrow x^2 + 2x + 3x + 6 \\
&\longrightarrow x^2 + 5x + 6
\end{aligned}$$

You can ask Maple to leave sub-expressions intact, by specifying a sub-expression as an optional second argument
```
> p:=(x+2)*(x+3)^2:
> expand(p,x+3);
```

$$(x + 3)^2 x + 2 (x + 3)^2$$

```
> expand(p,x+2);
```

$$(x + 2) x^2 + 6 (x + 2) x + 9x + 18$$

Note that the term $9(x + 2)$ is automatically expanded.

The function `expand` also distributes positive integer powers over sums
```
> (x+1)^3;
```

$$(x + 1)^3$$

```
> expand(%);
```

$$x^3 + 3x^2 + 3x + 1$$

but negative powers are left untouched

```
> g:=(x+1)^(-2);
```

$$g := \frac{1}{(x + 1)^2}$$

```
> expand(g);
```

$$\frac{1}{(x + 1)^2}$$

In this case, the expansion of the denominator must be carried out explicitly

```
> 1/expand(denom(g));
```

$$\frac{1}{x^2 + 2x + 1}$$

The following example illustrates the different behaviour of expand on the numerator and the denominator of a rational expression

```
> x*(x+1)/(x-1)^2;
```

$$\frac{x (x + 1)}{(x - 1)^2}$$

```
> expand(%);
```

$$\frac{x^2}{(x - 1)^2} + \frac{x}{(x - 1)^2}$$

The procedure simplify provides — among other things — a standard simplification of rational functions over $\mathbb{Q}$. It transforms a rational function in the form

$$\frac{numerator}{denominator}$$

with numerator and denominator relatively prime polynomials with integer coefficients.

```
> x*(x+2/3*x^3)/(x+1/7);
```

$$\frac{x \left(x + \frac{2 x^3}{3} \right)}{x + \frac{1}{7}}$$

```
> simplify(%);
```

$$\frac{7}{3} \frac{x^2 \left(3 + 2 x^2 \right)}{7 x + 1}$$

In the above expression, all fractional coefficients have been cleared.

```
> x+1/(x^2+1/(x^3+1/x^4));
```

$$x + \cfrac{1}{x^2 + \cfrac{1}{x^3 + \cfrac{1}{x^4}}}$$

```
> simplify(%),factor(%);
```

$$\frac{x^{10} + x^3 + x^5 + x^7 + 1}{x^2 \left(x^7 + 1 + x^2\right)}, \quad \frac{x^{10} + x^3 + x^5 + x^7 + 1}{x^2 \left(x^2 + x + 1\right) \left(x^5 - x^4 + x^2 - x + 1\right)}$$

Here `simplify` has produced a partially factored denominator. By contrast, the function `factor`, when applied to a rational expression, provides a simplified and fully factored form, and may be used as an alternative to `simplify`, keeping in mind that it is more expensive to evaluate. For simplification of rational expressions, see also `?normal`.

A trick of the trade

We consider the problem of proving the following identity with Maple

$$(1 + x)^2 + \frac{1}{1 + x} = \frac{(1 + x)^3 + 1}{1 + x}.$$

```
> rf:=(1+x)^2+1/(1+x):
```

The procedure `simplify` does not do what we want

```
> simplify(rf);
```

$$\frac{2 + 3x + 3x^2 + x^3}{1 + x}$$

and factorization does not help, either

```
> factor(%);
```

$$\frac{(x + 2)(x^2 + x + 1)}{1 + x}$$

The idea is to perform a substitution in order to 'freeze' the subexpression $1 + x$.

```
> subs(1+x=y,rf);
```

$$y^2 + \frac{1}{y}$$

```
> simplify(%);
```

$$\frac{y^3 + 1}{y}$$

```
> subs(y=1+x,%);
```

$$\frac{(1 + x)^3 + 1}{1 + x}$$

7.5   Partial fractions decomposition*

The decomposition into partial fractions is an important representation of a rational function. We consider an example.

```
> (x+1)/(x^4-2*x^3+x^2-2*x);
```

$$\frac{x+1}{x^4 - 2x^3 + x^2 - 2x} \tag{7.4}$$

The conversion to partial fractions is done by the many-purpose intrinsic function `convert` (check the online documentation)

```
> convert(%,'parfrac',x);
```

$$-\frac{1}{2}\frac{1}{x} + \frac{3}{10}\frac{1}{x-2} + \frac{1}{5}\frac{-3+x}{x^2+1}$$

How does this decomposition work? The denominator of the rational function (7.4) factors into the product of three irreducible polynomials

$$\frac{x+1}{x^4 - 2x^3 + x^2 - 2x} = \frac{x+1}{x(x-2)(x^2+1)}.$$

The right-hand side is then converted to the form

$$\frac{q_1}{x} + \frac{q_2}{x-2} + \frac{q_3}{x^2+1}$$

where

$$q_1 = -\frac{1}{2} \qquad q_2 = \frac{3}{10} \qquad q_3 = -\frac{3}{5} + \frac{x}{5}.$$

The three polynomials q_k are polynomials over $\mathbb{Q}$ with degree lower than that of the corresponding denominator. The following theorem describes the general situation.

Theorem 9 *Let f/g be a rational function over $\mathbb{Q}$, whose numerator is of lower degree than the denominator. Let*

$$g = p_1^{e_1} \cdot p_2^{e_2} \cdots p_r^{e_r}$$

be the factorization of the denominator into irreducibles. Then there exist polynomials $q_1, \ldots q_r$, with $deg(q_k) < deg(p_k)$, for $k = 1, \ldots, r$, such that

$$\frac{f}{g} = \frac{q_1}{p_1^{t_1}} + \frac{q_2}{p_2^{t_2}} + \cdots + \frac{q_r}{p_r^{t_r}}$$

and $t_k \leq e_k$, $k = 1, \ldots, r$.

A common use of the decomposition into partial fractions is to integrate rational functions.

Exercises

Exercise 7.10. Consider the polynomial $(2x^3 - x)(x^2 - x - 1)$. Using Maple, transform this polynomial into:

(a) $\quad 2\,x^5 - 2\,x^4 - 3\,x^3 + x^2 + x$

(b) $\quad x\,(2x^2 - 1)(x^2 - x - 1)$

(c) $\quad x^3\,(2\,x^2 - 1) - x^2\,(2\,x^2 - 1) - x\,(2\,x^2 - 1)$

(d) $\quad 2\,x^3\,\left(x^2 - x - 1\right) - x\,\left(x^2 - x - 1\right)$

(e) $\quad x\,\left(2\,x^4 - 2\,x^3 - 3\,x^2 + x + 1\right).$

Exercise 7.11. Consider the rational function
$$\frac{x^3 - 3x + 2}{x^3 - 2x^2 - x + 2}.$$
Using Maple, transform this rational function into:

(a) $\quad \dfrac{(x-1)^2\,(x+2)}{x^3 - 2\,x^2 - x + 2}$

(b) $\quad \dfrac{x^3 - 3\,x + 2}{(x+1)\,(x-1)\,(x-2)}$

(c) $\quad \dfrac{(x-1)^2\,x + 2\,(x-1)^2}{(x-2)\,x^2 - x + 2}$

(d) $\quad \dfrac{x^2 + x - 2}{x^2 - x - 2}$

(e) $\quad \dfrac{(x+2)\,(x-1)}{(x+1)\,(x-2)}$

(f) $\quad \dfrac{x^2}{(x+1)\,(x-2)} + \dfrac{x}{(x+1)\,(x-2)} - \dfrac{2}{(x+1)\,(x-2)}.$

Exercise 7.12.

(a) Using Maple, show that
$$\frac{(x+1)^{1000} - 1}{(x+1)^{1000}} + \frac{1}{(x+1)^{999}} = \frac{x}{(x+1)^{1000}} + 1.$$

WARNING: make sure you do not expand the binomials!

(b) Using Maple, transform

$$x^2 - 2x + 1 + \frac{1}{x^2 - 2x + 1}$$

into

$$\frac{1 + (x-1)^4}{(x-1)^2}$$

and vice versa. (*Hint:* you may need asubs.)

Exercise 7.13.
(a) Show that $x^4 + x^3 + x^2 + x + 1$ and $x^2 + x + 1$ are relatively prime.
(b) List *all* common divisors of $x^6 - x^5 - 2x^4 + x^3 + x^2 + 2x - 2$ and $x^5 - x^3 - x^2 + 1$ in $\mathbb{Z}[x]$.
(c) Let

$$f(x) = x^6 - x - 1 \qquad g(x) = x^3 + x + 1.$$

Let $q(x)$ and $r(x)$ be the quotient and remainder, respectively, of the division of $f(x)$ by $g(x)$. Verify that

$$\frac{f(x)}{g(x)} = q(x) + \frac{r(x)}{g(x)}.$$

Exercise 7.14.
(a) Consider the *recursive* sequence of polynomials in $\mathbb{Z}[x]$

$$g_0(x) = x \qquad g_{t+1}(x) = 1 + x \cdot g_t(x), \qquad t \geq 0.$$

Determine the smallest value of t for which $g_t(x)$ is reducible.
(b) Consider the recursive sequence of rational functions over $\mathbb{Q}$

$$r_0(x) = x^3 \qquad r_{t+1}(x) = x + \frac{1}{r_t(x)}, \qquad t \geq 0.$$

Show that the numerator of $r_2(x)$ is divisible by $x^2 - x + 1$.

Exercise 7.15.   Let $h(x) = f(x)/g(x)$ be a rational function and let $r(x)$ be the remainder of division of $f(x)$ by $g(x)$. We define

$$\{h(x)\} = \frac{r(x)}{g(x)}.$$

Consider the function

$$\tau(h(x)) = \begin{cases} \left\{\frac{1}{h(x)}\right\} & h(x) \neq 0 \\ 0 & \text{otherwise.} \end{cases}$$

Show that the elements of the following recursive sequence of rational functions over $\mathbb{Q}$

$$h_0(x) = \frac{x^3 + x + 1}{x^5 + x + 1} \qquad h_{t+1}(x) = \tau(h_t(x)), \qquad t \geq 0$$

becomes eventually zero.

Chapter 8

Finite sums and products

In this chapter, we develop the tools for summing and multiplying a large (albeit finite) number of elements of a sequence. In this way, we shall construct new sequences from old ones.

8.1 Basics

We consider a sequence of elements of a set A

$$a_0, a_1, a_2, \ldots \qquad a_k \in A. \qquad (8.1)$$

All we require from A is that its elements can be added and multiplied together, e.g., $A = \mathbb{N}, \mathbb{Z}, \mathbb{Q}[x], \ldots$ etc. Then we form sums and products of the elements of the sequence (8.1)

$$S_n = a_0 + a_1 + a_2 + \cdots + a_n = \sum_{k=0}^{n} a_k \qquad n \geq 0, \qquad (8.2)$$

$$P_n = a_0 \cdot a_1 \cdot a_2 \cdots a_n = \prod_{k=0}^{n} a_k \qquad n \geq 0. \qquad (8.3)$$

It should be clear that, for any given sequence a_k, the quantities S_n and P_n are functions of n, and hence they are new sequences

n	0	1	2	3	$\cdots$
a_n	a_0	a_1	a_2	a_3	$\cdots$
S_n	a_0	$a_0 + a_1$	$a_0 + a_1 + a_2$	$a_0 + a_1 + a_2 + a_3$	$\cdots$
P_n	a_0	$a_0 \cdot a_1$	$a_0 \cdot a_1 \cdot a_2$	$a_0 \cdot a_1 \cdot a_2 \cdot a_3$	$\cdots$

We begin with some examples.

Example 8.1. $A = \mathbb{Z}$, $a_n = 1$, $n \geq 0$.

n	0	1	2	3	4	5	$\cdots$
a_n	1	1	1	1	1	1	$\cdots$
S_n	1	2	3	4	5	6	$\cdots$
P_n	1	1	1	1	1	1	$\cdots$

We have $S_n = n + 1$, and $P_n = 1$.

Example 8.2. $A = \mathbb{Z}$, $a_n = n$, $n \geq 1$.

n	1	2	3	4	5	$\cdots$
a_n	1	2	3	4	5	$\cdots$
S_n	1	3	6	10	15	$\cdots$
P_n	1	2	6	24	120	$\cdots$

We have

$$S_n = \frac{n(n+1)}{2} \qquad P_n = n!\,. \tag{8.4}$$

The leftmost formula is the sum of the *arithmetic progression*.

Example 8.3. $A = \mathbb{Z}$, $a_n = (-1)^n$, $n \geq 0$.

n	0	1	2	3	4	5	$\cdots$
a_n	1	-1	1	-1	1	-1	$\cdots$
S_n	1	0	1	0	1	0	$\cdots$
P_n	1	-1	-1	1	1	-1	$\cdots$

We have

$$S_n = \frac{1 + (-1)^n}{2} \qquad P_n = (-1)^{\lfloor (n+1)/2 \rfloor}$$

where $\lfloor \cdot \rfloor$ is the floor function.

Example 8.4. $A = \mathbb{Z}[x]$, $a_n = x^n$, $n \geq 0$.

n	0	1	2	3	$\cdots$
a_n	1	x	x^2	x^3	$\cdots$
S_n	1	$1 + x$	$1 + x + x^2$	$1 + x + x^2 + x^3$	$\cdots$
P_n	1	x	x^3	x^6	$\cdots$

We have

$$S_n = \frac{x^{n+1} - 1}{x - 1} \qquad P_n = x^{n(n+1)/2}. \tag{8.5}$$

In the leftmost expression we can substitute for x any numerical value $x \neq 1$, to obtain the formula for the sum of the *geometric progression*.

Example 8.5. *Summing characteristic functions.* Let A be a subset of $\mathbb{N}$, and let χ_A be the characteristic function of A in $\mathbb{N}$ (cf. (3.14)). We wish to count the elements of A that are not greater than a given natural number n. Let this number be $A(n)$. Because $\chi_A(k)$ is 0 if $k \notin A$, and 1 if $k \in A$, we have

$$A(n) = \sum_{k=0}^{n} \chi_A(k).$$

For instance, if A is the set of primes, then $A(n)$ is the number of primes not exceeding n.

8.2 Sums and products with Maple

Whenever the elements of a sequence can be represented explicitly by a Maple expression, then any finite sum or product of these elements can be performed using the standard library functions `add` and `mul`. Sums and products of the type (8.2) and (8.3), respectively, are represented as

```
> add(expr, k=m..n);
```

```
> mul(expr, k=m..n);
```

where the first argument `expr` is any Maple expression representing the coefficients a_k, and the summation (multiplication) index k ranges between the numerical values m and n (usually integer). The syntax of `add` and `mul` is similar to that of `seq`. In particular, the range of summation (multiplication) can be replaced by any expression e

```
> add(expr, k=e);
```

```
> mul(expr, k=e);
```

in which case the sum (product) is performed evaluating a at each operand of e (see section 6.3).

Example 8.6. Compute the sum of the cubes of the first 100 positive integers. This is the sum of the first 100 terms of the sequence $a_k = k^3$, $k = 1, 2, \ldots$

$$\sum_{k=1}^{100} a_k \qquad a_k = k^3.$$

We first construct a user-defined function representing the summands a_k, and then perform the sum using such function

```
> cube:=k->k^3:
```

```
> add(cube(k), k=1..100);
```

$$25502500$$

Alternatively, the expression for the summands could have been stored in a variable, or used directly as the first argument of the function add

```
> a:=k^3:
```
```
> add(a,k=1..100),add(k^3, k=1..100);
```

$$25502500, \ 25502500$$

We shall, however, favour the explicit representation of the summands as values of a function, because this construct helps clarify the structure of the summation process.

Example 8.7. Compute the product of the first 20 prime numbers. The elements of the sequence of primes are available through the standard library function ithprime

```
> mul(ithprime(s),s=1..20);
```

$$557940830126698960967415390$$

Example 8.8. Verify that 33 factorial is actually the product of the first 33 positive integers

```
> evalb(mul(things,things=1..33)=33!);
```

$$true$$

In this example, the relevant sequence is just the identity sequence $a_k = k$, for which we have not constructed a user-defined function. The following (rather pedantic) alternative solution underlines this point

```
> id:=x->x:
```
```
> mul(id(i),i=1..33)-33!;
```

$$0$$

Example 8.9. We construct the function $\pi_1(n)$, which gives the number of primes of the form $4k + 1$ which are not greater than n. (This is a specialization of the prime counting function π introduced in section 2.6.) We first construct the characteristic function of the set of such primes, and then sum.

```
> chi:=k->if irem(k,4)=1 and isprime(k) then 1 else 0 fi:
```
```
> pi1:=n->add(chi(k),k=1..n):
```
```
> pi1(100);
```

$$11$$

Example 8.10. Consider the following sequence of elements of $\mathbb{Z}[x]$

$$S_n = \sum_{k=1}^{n} \left(x^k - kx + 1 \right) \qquad\qquad n \geq 1.$$

We show that the polynomial S_5 is irreducible.

```
> p:=k->x^k-k*x+1:
> S:=n->add(p(k),k=1..n):
> factor(S(5));
```

$$5 + x^2 - 14\,x + x^3 + x^4 + x^5$$

Example 8.11. *Partial factorizations.* We consider the polynomial identities

$$\sum_{k=0}^{11} x^k = (x+1)\,(x^2+1)\,(x^2+x+1)\,(1-x+x^2)\,(x^4-x^2+1)$$

$$= (x^2+x+1)\,(x^3+x^9+x^6+1).$$

How can we achieve the rightmost representation (a partial factorization), starting from the leftmost one (a full expansion)? We first need a full factorization

```
> p:=factor(add(x^k,k=0..11));
```

$$p := (x+1)\,(x^2+1)\,(x^2+x+1)\,(1-x+x^2)\,(x^4-x^2+1)$$

This is an expression of the '*' type, with 5 operands

```
> whattype(p),nops(p);
```

$$*, 5$$

To achieve a partial factorization, we must assemble and multiply together the relevant operands

```
> op(3,p)*expand(mul(op(k,p),k=[1,2,4,5]));
```

$$(x^2+x+1)\,(x^3+x^9+x^6+1)$$

The same construct is also possible with integers

```
> ifactor(30!);
```

$$(2)^{26}\,(3)^{14}\,(5)^7\,(7)^4\,(11)^2\,(13)^2\,(17)\,(19)\,(23)\,(29)$$

```
> op(1,")*op(nops("),")*expand(mul(op(k,"),k=2..nops(")-1));
```

$$136295711075535652265625\,(2)^{26}\,(29)$$

(It is worth recalling that the order in which the factors are displayed is context-dependent.)

8.3   Symbolic evaluation of sums and products

If the elements a_k of a sequence are expressed as an explicit function of k, can the associated sum S_n or product P_n be expressed as an explicit function of n? This is clearly possible in some cases, e.g., for the sum of the arithmetic and geometric progressions (see (8.4) and (8.5))

$$\sum_{k=1}^{n} k = \frac{n(n+1)}{2} \qquad \sum_{k=0}^{n} x^k = \frac{x^{n+1}-1}{x-1}, \qquad x \neq 1. \qquad (8.6)$$

Such formulae allow us to compute the above sums for any given value of n, with much less computational work than carrying out the sum explicitly

$$\sum_{k=10^3}^{10^4} k = \frac{10^4(10^4+1)}{2} - \frac{10^3(10^3+1)}{2} = 49505500. \qquad (8.7)$$

Note the structural analogy with the problem of constructing indefinite integrals

$$\int x\,dx = \int_0^x y\,dy = \frac{1}{2}x^2.$$

Maple provides the standard library functions sum and product which are designed to perform symbolic (i.e., indefinite) summations and products, rather than to explicitly add or multiply quantities. Their syntax is the same as that of add and mul, except that the range of summation (product) may now be symbolic, not just numerical. The function sum will attempt to 'discover' a formula for indefinite summation every time it is called with a non-numerical range (the reader should be aware that this process is time-consuming). If the range is numerical, sum and product will give the same result as add and mul, respectively, *except a lot more slowly*. So the use of these functions for definite summations is to be avoided.

For instance, sum can find the formula for the indefinite sum of the arithmetic and geometric progressions (8.6)

```
> AS:=sum(k,k=1..n);
```

$$AS := \frac{1}{2}(n+1)^2 - \frac{1}{2}n - \frac{1}{2}$$

```
> factor(%);
```

$$\frac{1}{2}n(n+1)$$

The function factor is legitimate here, because the value of the sum is a polynomial in $\mathbb{Q}[n]$. Now that we have a formula, we can compute the sum (8.7) with substitutions

```
> subs(n=10^4,AS)-subs(n=10^3,AS);
```

$$49505500$$

The geometric sum is treated similarly

```
> GS:=sum(x^k,k=1..n);
```

$$GS := \frac{x^{(n+1)}}{x-1} - \frac{1}{x-1}$$

```
> simplify(%);
```

$$\frac{x^{(n+1)} - 1}{x - 1}$$

The reader is invited to compare the amount of computation involved in evaluating the arithmetic sum for a large value of n, using **add**, **sum** with a numeric argument, and substitution into the symbolic expression **AS**.

Let us consider another example. The terms of the sum

$$S_n = \sum_{k=1}^{n} \binom{k^2}{k}$$

are given by an explicit function of k. Is there a simple formula for S_n this time?

```
> bin:=k->binomial(k^2,k):
> S:=sum(bin(k),k=1..n):
> S;
```

$$\sum_{k=1}^{n} binomial(k^2, k)$$

Maple cannot find such a formula. This, of course, does not mean that there isn't one, but only that the function **sum** cannot find it. Substituting an integer value for n in the expression S will cause the definite summation to be performed explicitly

```
> subs(n=4,S);
```

$$1911$$

However, since the evaluation of the indefinite sum failed in this case, the use of **add** would be preferable.

The functions **sum** and **product** have the *inert form* **Sum** and **Product**, respectively, which leaves the operation indicated. The function **value** will turn the value of the inert form into the actual value.

```
> Sum(cube(k),k=1..100);
```

$$\sum_{k=1}^{100} k^3$$

```
> value(%);# eval does not work here
```
$$25502500$$

The inert form can be combined with `value` to obtain a nice layout
```
> Sum(cube(k),k=1..100):%=value(%);
```
$$\sum_{k=1}^{100} k^3 = 25502500$$

(For further applications of these inert functions, see the next section.)

8.4   Double sums and products

A *double sum* is an expression of the type

$$S = S_{m,n} = \sum_{j=m_0}^{m} \sum_{k=n_0}^{n} a_{j,k} = \sum_{j=m_0}^{m} \left(\sum_{k=n_0}^{n} a_{j,k} \right) \qquad (8.8)$$

where $a_{j,k}$ is a function of both j and k. The sum S depends also on n_0 and m_0, but for the time being we do not make this dependence explicit.

A double sum can be thought of as a sum of elements of a sequence which are themselves sums

$$S_{m,n} = \sum_{j=m_0}^{m} s_{j,n} \qquad s_{j,n} = \sum_{k=n_0}^{n} a_{j,k}.$$

The dependence of $s_{j,n}$ on j not only derives from $a_{j,k}$, but it may also result from the j-dependence of the range $n_0 \ldots n$. In other words, we may have $n_0 = n_0(j)$ and $n = n(j)$

$$\sum_{j=m_0}^{m} \sum_{k=j}^{j^2} a_{j,k}.$$

The simplest case is when the range $n_0 \ldots n$ of the inner sum is independent from the summation index of the outer sum. This is illustrated in the following example

$$\begin{aligned} S = \sum_{j=1}^{3} \sum_{k=1}^{2} a_{j,k} &= \sum_{j=1}^{3} \left(\sum_{k=1}^{2} a_{j,k} \right) \\ &= \sum_{j=1}^{3} \left(a_{j,1} + a_{j,2} \right) \\ &= a_{1,1} + a_{1,2} + a_{2,1} + a_{2,2} + a_{3,1} + a_{3,2}. \end{aligned}$$

The sum S can be written as

$$S = \sum_{j=1}^{3} S_j \qquad S_j = \sum_{k=1}^{2} a_{j,k} = a_{j,1} + a_{j,2},$$

showing that S is the sum of the first three elements of the sequence S_j, which are themselves sums. Because the upper index of the inner sum does not depend on the summation index of the outer sum, the order of summation can be reversed

$$\begin{aligned}
\sum_{j=1}^{3} \sum_{k=1}^{2} a_{j,k} &= \sum_{k=1}^{2} \sum_{j=1}^{3} a_{j,k} \\
&= \sum_{k=1}^{2} (a_{1,k} + a_{2,k} + a_{3,k}) \\
&= a_{1,1} + a_{2,1} + a_{3,1} + a_{1,2} + a_{2,2} + a_{3,2}.
\end{aligned}$$

The innermost range does, in general, depend on the outer index. In the following example, we have $n = n(j) = j$

$$\begin{aligned}
\sum_{j=1}^{3} \sum_{k=1}^{j} a_{j,k} &= \sum_{j=1}^{3} \left(\sum_{k=1}^{j} a_{j,k} \right) \\
&= \sum_{k=1}^{1} a_{1,k} + \sum_{k=1}^{2} a_{2,k} + \sum_{k=1}^{3} a_{3,k} \\
&= a_{1,1} + a_{2,1} + a_{2,2} + a_{3,1} + a_{3,2} + a_{3,3}.
\end{aligned}$$

In this case, the order of summation cannot be reversed, for otherwise the range of summation of the k-sum would be undefined.

Double products behave in the same way as double sums

$$\begin{aligned}
\prod_{j=1}^{3} \prod_{k=1}^{2} a_{j,k} &= \prod_{j=1}^{3} (a_{j,1} \cdot a_{j,2}) \\
&= a_{1,1} \cdot a_{1,2} \cdot a_{2,1} \cdot a_{2,2} \cdot a_{3,1} \cdot a_{3,2}.
\end{aligned}$$

As with sums, when the ranges are independent, the order of multiplication can be reversed.

Sums and products can be combined

$$\begin{aligned}
\sum_{j=1}^{3} \left(\prod_{k=1}^{2} a_{j,k} \right) &= \sum_{j=1}^{3} (a_{j,1} \cdot a_{j,2}) \\
&= a_{1,1} \cdot a_{1,2} + a_{2,1} \cdot a_{2,2} + a_{3,1} \cdot a_{3,2}.
\end{aligned}$$

Interchanging $\sum$ and $\prod$ alters completely the meaning of the expression

$$\begin{aligned}
\prod_{k=1}^{2} \left(\sum_{j=1}^{3} a_{j,k} \right) &= \prod_{k=1}^{2} (a_{1,k} + a_{2,k} + a_{3,k}) \\
&= (a_{1,1} + a_{2,1} + a_{3,1}) \cdot (a_{1,2} + a_{2,2} + a_{3,2}) \\
&= a_{1,1} \cdot a_{1,2} + a_{1,1} \cdot a_{2,2} + a_{1,1} \cdot a_{3,2} \\
&\quad + a_{2,1} \cdot a_{1,2} + a_{2,1} \cdot a_{2,2} + a_{2,1} \cdot a_{3,2} \\
&\quad + a_{3,1} \cdot a_{1,2} + a_{3,1} \cdot a_{2,2} + a_{3,1} \cdot a_{3,2}.
\end{aligned}$$

Double sums and products with Maple

Consider the expression

$$S = \prod_{j=1}^{4} \left(\sum_{k=1}^{j} (j + k)^2 \right) = \prod_{j=1}^{4} s_j.$$

In the notation of (8.8), we have $a_{j,k} = (j + k)^2$, so we begin writing a user-defined function for a

```
> a:=(j,k)->(j+k)^2:
```

Next we write the inner sum as a function, using the inert summation

```
> s:=j->Sum(a(j,k),k=1..j):
```

and finally we write out the product

```
> Product(s(j),j=1..4):%=value(%);
```

$$\prod_{j=1}^{4} \left(\sum_{k=1}^{j} (j + k)^2 \right) = 1339800$$

Combining the active function **product** (or, better still, **mul**) with the inert function **Sum**, the product is evaluated but the sum is not, giving

```
> product(s(j),j=1..4);
```

$$\left(\sum_{k=1}^{1} (1 + k)^2 \right) \left(\sum_{k=1}^{2} (2 + k)^2 \right) \left(\sum_{k=1}^{3} (3 + k)^2 \right) \left(\sum_{k=1}^{4} (4 + k)^2 \right)$$

The following statements extract and evaluate the third operand in the above expression

```
> op(3,%):%=value(%);
```

$$\sum_{k=1}^{3} (3 + k)^2 = 77$$

Example 8.12. We wish to count the *visible points* (see figure 6.5, section 6.3), that lie within a square region on the plane, centered at the origin. This will involve performing a double summation of their characteristic function. Because visible points are characterized by the fact that their coordinates are coprime (theorem 6 of section 6.3), we deduce that the pattern of visible points in the plane must posses an 8-fold symmetry, since for any integers x and y we have $\gcd(\pm x, \pm y) = \gcd(\pm y, \pm x)$. We denote by $Vp(n)$ the total number of visible points lying within the square $-n \leq x, y, \leq n$, where n is a positive integer. One sees that $Vp(1) = 8$. For $n \geq 2$, we must add to this figure eight times the number of the visible points contained within the triangular region defined by the inequalities

$$2 \leq x \leq n \qquad\qquad y \leq x - 1. \qquad\qquad (8.9)$$

Such inequalities determine the range of double summation of the characteristic function chi(x,y).

```
> chi:=(x,y)->if igcd(x,y)=1 then 1 else 0 fi:
> Vp:=n->8*(1+add(add(chi(x,y),y=1..n-1),x=2..n)):
> Vp(1),Vp(10),Vp(100),Vp(1000);
```

$$8,\ 256,\ 24352,\ 2433536$$

For $n = 1$, the function Vp gives the correct answer 8, since in this case, the outer function add has an empty range of summation: (x=2..1).

What is the probability that an integer point chosen at random is visible? From Theorem 6, this is the same as the chance that two randomly chosen integers are coprime, and a theorem in number theory states that the latter probability is equal to $6/\pi^2$. We estimate it by dividing $Vp(n)$ by the total number of integer points in the corresponding box, which is equal to $(2n)^2$. For this purpose, we use the largest value computed above ($n = 1000$), corresponding to a sample of 4 million points

```
> evalf(2433536/(2*1000)^2),evalf(6/Pi^2);
```

$$.6083840000, .6079271016$$

Our result is accurate to within three decimal figures.

Exercises

Exercise 8.1. Compute the value of the following expressions (p_k denotes the kth prime number).

(a) $\displaystyle\sum_{k=0}^{100} k(k+1)$

(b) $\displaystyle\sum_{k=0}^{4} \frac{k^2 - 10}{k^2 + k + 1}$

(c) $\displaystyle\prod_{k=1}^{10} \left(1 - \frac{1}{p_k}\right)^{-1}$

(d) $\displaystyle\sum_{k=0}^{50} (-1)^k \binom{50}{k}$

(e) $\displaystyle\sum_{n=0}^{5} \frac{x^n}{n!}$

(f) $\displaystyle\sum_{n=1}^{5} (-1)^{n+1} \frac{x^{2n-1}}{(2n-1)!}$

(g) $\displaystyle\sum_{k=0}^{10} \binom{\binom{k}{2}}{2} \binom{20-k}{10}$

(h) $\displaystyle\prod_{k=1}^{5} \left(\sum_{j=1}^{10} (k+j)\right)$

(i) $\displaystyle\sum_{k=0}^{5} \sum_{j=1}^{k+1} \frac{k^2 - j}{k + j}$

(j) $\displaystyle\sum_{n=1}^{5} \sum_{k=0}^{n} \frac{2^{k+1}}{k + 1} \binom{n}{k}.$

Exercise 8.2. Using Maple, prove that

$$\prod_{k=1}^{4} \left(\sum_{j=1}^{k+1} (k^2 + j)\right)^{k-2} = \frac{\left(\displaystyle\sum_{j=1}^{4} (9 + j)\right) \left(\displaystyle\sum_{j=1}^{5} (16 + j)\right)^2}{\displaystyle\sum_{j=1}^{2} (1 + j)}.$$

Exercise 8.3.   Using Maple, prove that

(a) $\displaystyle\sum_{k=1}^{n} k^2 = \frac{1}{6} n(n+1)(2n+1)$ (b) $\displaystyle\sum_{k=1}^{n} \frac{1}{k(k+1)} = \frac{n}{n+1}$

(c) $\displaystyle\sum_{k=1}^{n} k\, 2^k = (n-1)\, 2^{n+1} + 2$ (d) $\displaystyle\prod_{n=2}^{m} \left(1 - \frac{1}{n^2}\right) = \frac{1}{2}\frac{m+1}{m}$

(e) $\displaystyle\sum_{k=1}^{n} k^3 = \left(\sum_{k=1}^{n} k\right)^2$ (f) $\displaystyle\sum_{k=1}^{n} k^2 \binom{n}{k} = 2^{n-2} n\,(n+1).$

Exercise 8.4.   Express each of the following sums (products) by means of the summation (product) symbol. Then, using Maple, find and prove formulae for their values. Express the answer in the simplest possible form.

(a) $1 \cdot 1! + 2 \cdot 2! + 3 \cdot 3! + \quad \cdots \quad + (n-1) \cdot (n-1)!$

(b) $\dfrac{1}{1 \cdot 4} + \dfrac{1}{4 \cdot 7} + \dfrac{1}{7 \cdot 10} + \quad \cdots \quad + \dfrac{1}{(3n-2) \cdot (3n+1)}$

(c) $\left(1 - \dfrac{4}{1}\right)\left(1 - \dfrac{4}{9}\right)\left(1 - \dfrac{4}{25}\right) \quad \cdots \quad \left(1 - \dfrac{4}{(2n-1)^2}\right).$

Exercise 8.5.   The purpose of this exercise is to carry out computer-assisted proofs by induction of summation and product formulae. We illustrate the procedure for summations — which can be easily generalized to the case of products. We consider expressions of the form

$$\sum_{k=n_0}^{n} a_k = F(n) \qquad n \geq n_0 \qquad\qquad (8.10)$$

where n_0 is an integer and F is a given function of n. We define $S(n) = \sum_{k=n_0}^{n} a_k$ and rewrite equation (8.10) as

$$S(n) = F(n) \qquad n \geq n_0. \qquad\qquad (8.11)$$

To prove (8.11) you must proceed as follows:

- Construct Maple functions for the indefinite summation $S(n)$ and its conjectured value $F(n)$.

- Using these functions, verify that $S(n_0) = F(n_0)$, thereby establishing (8.11) for $n = n_0$.

- Assume that (8.11) is valid for some $n \geq n_0$ (this is the induction hypothesis), and hence express $S(n+1)$ as a function of $F(n)$

$$S(n+1) = S(n) + a_{n+1} = F(n) + a_{n+1}.$$

- Verify with Maple that $F(n) + a_{n+1} = F(n+1)$, proving that equation (8.11) holds for $n+1$.

Following the above procedure, prove the formulae

(a) $$\sum_{k=1}^{n} k^5 = \frac{1}{12} n^2 (2n^2 + 2n - 1)(n+1)^2 \quad n \geq 1$$

(b) $$\prod_{k=0}^{n-1} (1 + x^{2^k}) = \frac{1 - x^{2^n}}{1 - x} \quad n \geq 0$$

(c) $$\sum_{k=1}^{n} \frac{k^2 + k - 1}{(k+2)!} = \frac{1}{2} - \frac{n+1}{(n+2)!} \quad n \geq 1.$$

Exercise 8.6. By summing a suitable characteristic function, determine
(a) The number of composite integers between 900 and 1000 (endpoints included).
(b) The number of odd multiples of 31 lying between 10000 and 11000.
(c) The number of positive divisors of 16200 which are not greater than 150.
(d) The number of *integer* solutions n to the inequalities

$$n + 2 + (n+2)^3 > n^4 + 1 \qquad |n| < 10.$$

[*Hint:* interpret the rightmost inequality as a range of summation.]
(e) The number of points on the plane which have integer coordinates and which lie *inside* the circle with radius $\sqrt{10}$. How many points lie on the circumference?

Exercise 8.7*. The *Euler's totient function* $\phi(n)$ counts the number of positive integers that are smaller than n and relatively prime to it. There are 4 integers smaller than 8 and relatively prime to it, namely 1,3,5, and 7. Thus, $\phi(8) = 4$. Construct a user-defined function for $\phi(n)$, and hence determine an integer n such that $\phi(n)/n < 1/6$.
[*Hint:* plot the elements of the sequence $\phi(n)/n$. To speed up your computations, you may use Maple's version of ϕ, called `numtheory[phi](x)`, which is very fast.]

Exercise 8.8*. Let n be an integer. Construct a function `SumDigits(n)` whose value is the sum of the decimal digits of n.

8.5   Sums and products as recursive sequences

We consider again the problem of adding or multiplying the elements of a sequence $\{a_k\}$, in the case in which a_k is not given as an explicit function of the subscript k. This is the case, for instance, when a_k is defined recursively. Under these circumstances, the function **add** is of no use, and we must resort to recursive summation.

We first rewrite finite sums and products as follows (cf. equations (8.2) and (8.3))

$$S_n = \sum_{k=n_0}^{n} a_k = \sum_{k=n_0}^{n-1} a_k + a_n = S_{n-1} + a_n \qquad (8.12)$$

$$P_n = \prod_{k=n_0}^{n} a_k = \prod_{k=n_0}^{n-1} a_k \cdot a_n = P_{n-1} \cdot a_n. \qquad (8.13)$$

These are recursive formulae, valid for $n > n_0$. The leftmost summation and product are valid also for $n = n_0$, and substituting this value we obtain the initial conditions

$$\begin{aligned} S_{n_0} &= \textstyle\sum_{k=n_0}^{n_0} a_k = a_{n_0} \\ P_{n_0} &= \textstyle\prod_{k=n_0}^{n_0} a_k = a_{n_0}. \end{aligned}$$

The initial conditions can be made independent from the sequence $\{a_k\}$ as follows

$$S_{n_0} = S_{n_0-1} + a_{n_0} \quad \Longrightarrow \quad S_{n_0-1} = 0$$

$$P_{n_0} = P_{n_0-1} \cdot a_{n_0} \quad \Longrightarrow \quad P_{n_0-1} = 1,$$

extending the range of validity of the recursion formulae (8.12) and (8.13) to the case $n = n_0$

$$S_{n_0-1} = 0; \qquad S_n = S_{n-1} + a_n, \qquad n \ge n_0$$

$$P_{n_0-1} = 1; \qquad P_n = P_{n-1} \cdot a_n, \qquad n \ge n_0.$$

The last formulae yield the following recursive algorithm for computing sums and products

$$\begin{array}{llll}
S & \leftarrow & 0 & \qquad\qquad P \leftarrow 1 \qquad\qquad\qquad (8.14) \\
S & \leftarrow & S + a_{n_0} & \qquad\qquad P \leftarrow P \cdot a_{n_0} \\
S & \leftarrow & S + a_{n_0+1} & \qquad\qquad P \leftarrow P \cdot a_{n_0+1} \\
S & \leftarrow & S + a_{n_0+2} & \qquad\qquad P \leftarrow P \cdot a_{n_0+2} \\
& \vdots & & \qquad\qquad\qquad \vdots
\end{array}$$

Computing S_n or P_n requires iterating this procedure n times.

Example 8.13. *Summing a recursive sequence.* Let

$$a_0 = 2 \qquad a_{t+1} = f(a_t) = 1 - a_t^2, \qquad t \geq 0.$$

Compute

$$\sum_{t=0}^{5} a_t.$$

We resort to recursive summation, since we do not have an explicit expression for the summands a_t. We first construct a user-defined function for f

```
> f:=a->1-a^2:
```

Then we initialize the value of the sum, which is to be stored in the variable s.

```
> s:=0:
```

Next we initialize a, and begin summing the first term

```
> a:=2:s:=s+a:
```

(Clearly the above initialization could have been replaced by `a:=2:s:=a:`.) Then we update the value of a and add the letter to the sum, as many times as required.

```
> a:=f(a):s:=s+a:
> a:=f(a):s:=s+a:
> a:=f(a):s:=s+a:
> a:=f(a):s:=s+a:
> a:=f(a):s:=s+a:
```

Finally, we display the value of the sum

```
> s;
```

$$-15749063$$

Two comments are in order here. First, no subscript was necessary to carry out this computation, and second, the process of summation amounts to repetition of the same expression. In the next chapter, we will learn how to carry out this repetition automatically.

Exercises

Exercise 8.9. Consider the following recursive sequence in $\mathbb{Q}(x)$

$$g_0 = -x \qquad g_{t+1} = \frac{x}{g_t + 1}, \qquad t \geq 0.$$

Show that

$$\prod_{t=0}^{3} g_t = \frac{x^4}{x^2 - x - 1}.$$

Exercise 8.10. Let

$$h_0 = -1 \qquad h_1 = 1 \qquad h_{t+1} = 2\,h_t - h_{t-1}, \qquad t \geq 1.$$

Compute

$$\sum_{t=1}^{5} \frac{1}{h_t}.$$

Chapter 9

Elements of programming

In this chapter, we introduce three universal programming tools: iteration (the do-structure), conditional execution (the `if`-structure), and procedures (user-defined functions). Only the first one is new to us. We have encountered simple instances of the `if` structure in the context of characteristic functions, and we have used extensively the arrow operator `->` to construct user-defined functions consisting of a single statement.

9.1 Iteration

The `do`-structure allows repeated execution of a statement or a block of statements, which is one of the things computers are good at.

We illustrate the use of the `do`-structure for the construction of a recursive sequence. We wish to verify that the element α_3 of the recursive integer sequence

$$\alpha_0 = 1 \qquad \alpha_{t+1} = f(\alpha_t) = \alpha_t^5 + 1, \qquad t \geq 0,$$

has 8 decimal digits. As usual, we define the function f, initialize α, iterate as many times as required, and finally display the value of α

```
> f:=a->a^5+1:
> a:=1:
> a:=f(a):
> a:=f(a):
> a:=f(a):
> a >= 10^7 and a < 10^8;
```

true

The statement `a:=f(a):` is repeated three times. This process can be automated by means of the do-structure, as follows

163

```
> a:=1:
> to 3 do
>    a:=f(a)
> od:
> a;
```

$$39135394$$

The same result may be achieved with the following simpler construction, which makes use of ditto variables

```
> 1:
> to 3 do
>    f(%);
> od:

> %;
```

$$39135394$$

Next we consider the problem of constructing Pascal's triangle. Recall that the nth row of the triangle is given by the finite sequence

$$\binom{n}{0}, \binom{n}{1}, \ldots, \binom{n}{n}$$

which is best dealt with by a user-defined function

```
> row:=n->seq(binomial(n,k),k=0..n):
```

The following command generates the first eight rows of Pascal's triangle (n ranging from 0 to 7)

```
> row(0);row(1);row(2);row(3);row(4);row(5);row(6);row(7);
```

This time the statements to be repeated are not identical, but they depend on the index n increasing in unit steps from 0 to the prescribed maximal value. The do-structure can be made to assume the control of the integer variable n, as follows

```
> for n from 0 to 7 do
>    row(n)
> od;
```

$$1$$
$$1, 1$$
$$1, 2, 1$$
$$1, 3, 3, 1$$
$$1, 4, 6, 4, 1$$
$$1, 5, 10, 10, 5, 1$$
$$1, 6, 15, 20, 15, 6, 1$$
$$1, 7, 21, 35, 35, 21, 7, 1$$

The value of n is initialized to 0, and then increased by 1 each time. For every value of n the statement row(n) is executed, until n exceeds the target value of 7.

Syntax

The syntax of the do-structure is the following:
```
> for n from start by step to finish do
>     statement;
>     ...
>     statement
> od;
```
The *loop control variable* n is initialized at *start* and then increased by *step* until its value exceeds that of *finish* (or until it becomes smaller than *finish,* if *step* is negative). The expressions *start, step,* and *finish* must have integer, rational, or floating-point value.

The body of the do-structure consists of an arbitrary number of statements, each statement in the block being executed in correspondence of each value assumed by n. The loop control variable can be used within the structure as any other variable, but its value cannot be changed inside the loop. The last statement in the body of the structure does not require a terminator. To make the do-structure visible, it is good programming practice to *indent* the statements in the body of the loop, inserting a few blank characters.

The output of the *statements* is displayed if od is followed by a semicolon, and suppressed if od is followed by a colon, regardless of which terminator follows the individual statements.

The from and by options may be omitted, in which case the value of both *start* and *step* defaults to 1. If the loop control variable is not needed, the for option may be omitted, in which case Maple defines its own loop control variable. If the to option is omitted, the loop is infinite, and presumably it will be terminated by other means.

In the following examples, observe carefully the values assumed by the loop control variable on exit. Also note that the whole loop is logically equivalent to a single command.
```
> for i from 2 by 2 to 7 do i od:
```
$$2$$
$$4$$
$$6$$
```
> i;
```
$$8$$
```
> for i from 2 by -3 to -1 do i od;
```
$$2$$

$$-1$$

```
> i;
```

$$-4$$

If the range is empty, the loop is not executed, and the value of the loop control variable is not modified

```
> for i from 2 by -1 to 3 do i od;
> i;
```

$$2$$

To display the output of selected statements in a loop (rather than of all or no statements, as determined by the terminator of od), Maple provides the standard library function `print`, which always displays its value within the body of a do-structure, in the form of an expression sequence.

```
> x:=1:
> for n to 3 do
>     x:=x+igcd(n,x):
>     print(n,%)
> od:
```

$$1, 2$$
$$2, 4$$
$$3, 5$$

The colon after od suppresses all output, except that of the `print` statement, which is the only output to be shown.

Example 9.1.   An instructive comparison between two similar constructions using do and **seq** is as follows

```
> 2:
> for i to 3 do
>     %*i
> od;
```

$$2$$
$$4$$
$$12$$

```
> 2:
> seq(%*i,i=1..3);
```

$$2, 4, 6$$

Make sure you understand how and why these two results differ.

Example 9.2. *Summing the elements of a recursive sequence.* Compute the sum $\sum_{t=0}^{5} x_t$, with x_t given by (9.0), using the recursive summation algorithm (8.14).

```
> x:=0:
> s:=x:
> to 5 do
>    x:=f(x):
>    s:=s+x:
> od:
> s;
```

$$711$$

Example 9.3. *Constructing a nested expression.* Compute the value of the following expression

$$1^2 + \cfrac{1}{3^2 + \cfrac{1}{5^2 + \cfrac{1}{7^2 + \cfrac{1}{9^2}}}}$$

```
> 9^2:
> for i from 7 by -2 to 1 do
>    i^2+1/%
> od:
> %;
```

$$\frac{997280}{897949}$$

Example 9.4. *Plotting a recursive sequence.* We wish to plot the elements of the rational sequence

$$x_0 = 1/4 \qquad x_{t+1} = 1 - 2x_t^2, \qquad t \geq 0. \qquad (9.1)$$

As usual, we rewrite the recursion formula as $x_{t+1} = f(x_t)$, with $f(x) = 1 - 2x^2$. In the recursive computation of x_t, the index t plays no role, and the successive values of x_t are to be stored in the single variable x. The plotting routine requires a list $L = [[1, x_1], [2, x_2], \ldots]$, containing the points to be plotted, vs. the corresponding index. Such list L will be represented as an *expression sequence* of two-element lists, which we initialize to the NULL value. At the end of the computation, L is then turned into a list for plotting. The sequence (9.1) is rational, since x_0 is rational, and the recursion involves only rational operations. However, an inspection of the

first few elements reveals that numerator and denominator of x_t grow at
an alarming rate

$$x_1 = \frac{7}{8}, \qquad x_2 = \frac{-17}{32}, \qquad x_3 = \frac{223}{512}, \qquad x_4 = \frac{81343}{131072}.$$

Indeed, the number of digits of numerator and denominator of x_t approximately *doubles* at each step, due to the fact that f squares its argument
(the denominator of x_{10} has 309 digits!). So it would be impossible to plot,
say, the first 100 elements of the sequence (9.1), using an exact representation. We shall therefore opt for the *floating-point* representation of x_t
(see section 5.2), which is inexact, but which will render the computation
feasible, since it allocates a fixed amount of information for each x_t.

```
> f:=x->1-2*x^2:
> x:=0.25:
> L:=NULL:
> for t to 100 do
>    x:=f(x):
>    L:=L,[t,x]:
> od:
> L:=[L]:
> plot(L);
```

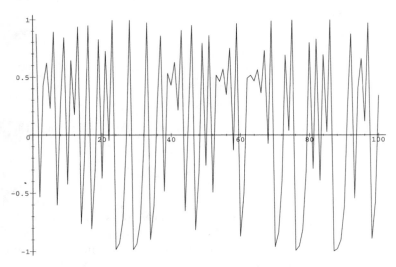

Note the very irregular behaviour of this sequence, which is *chaotic*, using
the jargon of the theory of dynamical systems [6].

The in *option*

Recall that functions such as `seq`, `add`, `mul`, etc., accept any expression as range, not just a sequence of integers.

```
> ifactor(3960);
```

$$(2)^3 \, (3)^2 \, (5) \, (11)$$

```
> [seq(expand(k),k=%)];
```

$$[8, 9, 5, 11]$$

A similar construction is available in a loop, using the in option

```
> for k in ifactor(3960) do
>    expand(k)
> od;
```

$$8$$
$$9$$
$$5$$
$$11$$

```
> L:=NULL:
> for k in ifactor(3960) do L:=[L,expand(k)] od:
> L;
```

$$[[[[8], 9], 5], 11]$$

The while *option*

The `while` option affords an alternative way to terminate an iteration. Its syntax is the following

```
> while logical expression do
>    statements
> od;
```

The logical expression in the heading is evaluated. If its value is *false*, the loop is exited. If it is *true*, the body of the loop is executed, after which the logical expression is evaluated again, etc. Evaluation does not require the use of `evalb`.

```
> x:=2:
> while x < 100 do
>    x:=x^2;
> od;
```

$$x := 4$$
$$x := 16$$
$$x := 256$$

The `while` option is the most general one, in the sense that the effect of any other combination of options can be reproduced using the `while` option alone. The `while` option can be used in conjunction with any other option. For instance, let `start`, `step`, and `finish` have integer values, and let `step` be positive. The structure

```
> for k from start by step to finish do
>     statements
> od:
```

can be reproduced as follows

```
> k:=start:
> while k <= finish do
>     statements;
>     k:=k+step
> od:
```

If `step` is negative, then the inequality tested by the `while` option should be reversed.

Example 9.5. Compute the number of primes less than 100.

```
> for n while ithprime(n) < 100 do od:
> n-1;
```

$$25$$

(Why is the required output $n - 1$ and not n?)

Example 9.6. Use the `while` option to compute O_b, the smallest positive integer n for which $n!$ is greater than b^n. (This function was considered in an exercise of chapter 4.) Here we compute the value of O_5.

```
> H:=(n,b)->n!<=b^n:
> for n while H(n,5) do od;
> n;
```

$$12$$

Note that the body of the loop is empty. All the work is done by the loop control functions. By *nesting* the above loop inside another loop that controls the value of b, we can compute several elements of the sequence O_b. These are the first 10 of them

```
> for b to 10 do
>     for n while H(n,b) do od:
>     print(b,n)
> od:
```

$$1, 2$$
$$2, 4$$
$$3, 7$$
$$4, 9$$
$$5, 12$$
$$6, 14$$

$$7, 17$$
$$8, 20$$
$$9, 22$$
$$10, 25$$

Exercises ————————————————————————————————

Exercise 9.1. In the following exercises you must use the do-structure.
(*a*) Display the square of the first 10 prime numbers. Hence do the same using the function `seq`.
(*b*) Compute the sum of the first 10 prime numbers. (When you are sure that your code is working, suppress all intermediate output within the loop.) Hence check your result with the function `add`.
(*c*) Compute the product of the first 10 prime numbers. Hence check your result with the function `mul`.
(*d*) Let

$$p(x) = x\,(3 + x\,(6 + x\,(9 + x\,(12 + x\,(15 + x))))).$$

Show that $p(x)/x$ is irreducible.
(*e*) Show that the element x_{20} of the following sequence of integers

$$x_0 = 3 \qquad x_1 = 5 \qquad x_{t+1} = f(x_t, x_{t-1}) = 3\,x_t - 2\,x_{t-1}, \qquad t \geq 1$$

is equal to $2^{21} + 1$.

Exercise 9.2. Consider the following recursive sequence of rationals

$$x_0 = 0 \qquad x_{t+1} = f(x_t) = \frac{x_t - 2}{x_t + 2}, \qquad t \geq 0. \qquad (9.2)$$

(*a*) Compute the element x_{10}, without displaying any intermediate output.
(*b*) Compute the first 10 elements of the sequence (9.2), displaying the value of t and x_t at each step.
(*c*) Compute the sum of the first 10 elements of the sequence (9.2).
(*d*) Plot the elements of the sequence (9.2) for t in the range $30, \ldots, 100$, connecting points with segments.

Exercise 9.3. In the following exercises you must use the `while` option of the do-structure.
(*a*) Display the positive cubes smaller than 500.
(*b*) Compute the smallest integer greater than 1 which is relatively prime to 9699690.
(*c*) Compute the smallest integer $n > 1$ for which $n^2 - 1$ is divisible by 87.
(*d*) Compute the smallest positive integer n for which $n^2 + n + 41$ is *not* prime.

(e) Compute the smallest positive n which is a multiple of 3, and for which $2^n > n^4$.

(f) Consider the following recursive sequence of complex numbers

$$z_0 = -2\,i \qquad z_{t+1} = z_t^2 + \frac{4}{3}\,i, \qquad t \ge 0.$$

Compute the smallest value of t for which $|z_t| > 10^5$.

Exercise 9.4. We consider the number q_n of composite integers following the nth prime p_n. Thus, $q_4 = 3$, because the fourth prime $p_4 = 7$ is followed by the 3 composite integers $8, 9, 10$. For $n > 1$, q_n is always odd (make sure you believe this).

(a) Compute the smallest prime p_n which is followed by at least 9 composite integers. Display also the corresponding value of n.

(b) Let a_t be the smallest value of n for which $q_n \ge t$. Thus, p_{a_t} is the smallest prime which is followed by at least t composite integers. (Think about it. Read the definition again. In the previous problem you have computed a_9 and p_{a_9}.) Compute and display t, a_t, and p_{a_t}, for all odd t smaller than 20 (excluding $t = 1$).

Exercise 9.5. Exponentials grow faster than any power. In this problem, we compute the integer $n > 1$ at which the exponential sequence 2^n overtakes the power n^e, where e is a given positive integer. We call this number Θ_e. Compute the first 10 elements of the sequence

$$\Theta_1, \Theta_2, \Theta_3, \ldots$$

Exercise 9.6*. Compute the largest integer less than 1000 that can be expressed as a sum of two squares (e.g., $5 = 1^2 + 2^2$). No intermediate output should be displayed.

Exercise 9.7*. Consider the set of natural numbers, the sum of whose decimal digits is divisible by 7

$$\{0, 7, 16, 25, 34, 43, 52, 61, 70, 106 \ldots\}.$$

Compute the number of elements of this set which are smaller than 1000.

Exercise 9.8*. We define an infinite recursive sequence $S = s_1, s_2, \ldots$ of elements of the set $\{1, -2\}$. We first let $S = 1, -2, 1$, and then we apply repeatedly the substitution: $S \mapsto S, 1, S, -2, S, 1, S$. After t substitutions, S will have $4^{t+1} - 1$ elements. We then define the sequence

$$\sigma_0 = 0 \qquad \sigma_n = \sum_{k=1}^{n} s_k, \qquad n \ge 1.$$

Plot the first 4^4 elements of the sequence $0, z_0, z_1, z_2, \ldots$, where z_n are complex numbers defined as

$$z_n = \sum_{k=0}^{n} e^{\pi i \sigma_k /3}.$$

Connect points with segments. (The result, shown below, is an approximation to a *fractal* called *Koch's snowflake*.)

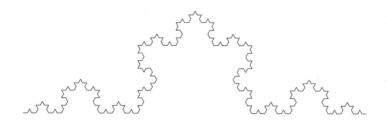

9.2 Study of an eventually periodic sequence

We now apply various techniques developed so far to the study of the structure of a family of eventually periodic sequences of integers, defined recursively.

Take a four-digit integer x_0, whose digits are not all the same: ($x_0 = 1066$). Rearrange the digits to make the largest possible number (6610) and the smallest (0166). Subtract the smallest from the largest ($6610 - 0166 = 6444 = x_1$). It can be shown (see exercises) that the result is a 4-digit integer whose digits are not all the same. By repeating the above procedure, we obtain a recursive sequence of four-digit integers: $x_{t+1} = F(x_t)$, $t \geq 0$.

The same construction can be carried out in the case of n-digit integers, for arbitrary $n > 1$

$$x_{t+1} = F_n(x_t) \qquad t \geq 0, \quad n > 1. \tag{9.3}$$

First some examples:

$$n = 2 : \quad 82, 54, 9, 81, 63, 27\ 45, 9, 81, \ldots$$
$$n = 4 : \quad 1066, 6444, 1998, 8082, 8532, 6174, 6174, \ldots$$

These sequences are *eventually periodic*, with period 5 and 1, respectively, and transient length 2 and 7, respectively. Eventual periodicity is easy to explain: the domain of F_n is *finite*, and therefore we must eventually encounter an integer that has appeared before. From that point on, the sequence will repeat.

We wish to explore the case $n = 4$ in full, producing a *computer-assisted proof* that all 4-digit sequences eventually reach 6174.

We need a user-defined function for F_n. Decimal digits are to be stored *backwards* in a 4-element list: 3861 <-> [1,6,8,3], and we must be able to switch easily from an integer to the corresponding list, and vice versa. For this purpose we develop two conversion functions. The function int2dig(x,n) produces the list of the first n digits of the integer x, beginning with the least significant one. If n is greater than the number of digits of x, the function will fill the missing spaces with zeros. This function is a variant of the function dgt developed in section 5.1 to compute the digits of a rational number.

```
> int2dig:=(x,n)->[seq(irem(iquo(x,10^(i-1)),10),i=1..n)]:
```
Its inverse of int2dig is dig2int(L), where L is a list of an arbitrary number of non-negative integers

```
> dig2int:=L->add(op(i,L)*10^(i-1),i=1..nops(L)):
```

Next we construct the largest and smallest n-digit integers obtainable from x. For this purpose, we use the standard library function sort, which, by default, sorts the elements of a list in *ascending order*

```
> big:=(x,n)->dig2int(sort(int2dig(x,n))):
```
To sort in descending order, we supply sort with an optional second argument, which is the symbolic name of a Boolean function $s(a, b)$ of *two* arguments, specifying the criterion for sorting. Specifically, $s(a, b)$ should return *true* if a is to precede b in the sorted list, and so the function for sorting in descending order is

```
> s:=(a,b)->not a < b:
```
whence

```
> small:=(x,n)->dig2int(sort(int2dig(x,n),s)):
```
As with map, the sorting function may be supplied directly, as a procedure definition

```
> small:=(x,n)->dig2int(sort(int2dig(x,n),(a,b)->not a < b)):
```
Finally, the desired function for the recursion (9.3) is

```
> F:=(x,n)->big(x,n)-small(x,n):
```

We now develop an algorithm to compute a given sequence up to and including its periodic part. Such a sequence will be stored in an expression sequence called S, to be initialized to NULL, the zero-length expression sequence. The variable NewS represents the sequence at the successive iteration. Periodicity is detected from the appearance of an element that already belongs to the sequence. To achieve this, the expression sequence is converted into a set, and each new element is checked for membership to that set.

```
> n:=4:
> x:=1066:
> S:=NULL:
> NewS:=x:
> while not member(x,{S}) do
>     S:=NewS:
>     x:=F(x,n):
>     NewS:=NewS,x
> od:
```

The variable S now contains all elements up to the first repetition, while the current value of x is that corresponding to the beginning of the repeating block. For convenience, we convert S into a list, and then identify the beginning of periodicity

```
> L:=[S]:
> for t while x <> op(t,L) do od:
```

Now the transient length is t-1. We display transient and repeating block using the *selection operator* (see section 6.1)

```
> L[1..t-1],L[t..nops(L)];
```

$$[1066, 6444, 1998, 8082, 8532], [6174]$$

We can now perform a global analysis of the sequence, by considering all possible 4-digit initial conditions. The latter will be stored in a set called ic. To avoid duplication, among the integers sharing the same digits, we consider only the smallest one. We must also eliminate integers with identical digits.

```
> ic:={seq(small(x,n),x=1..9998)} minus {i*1111$i=1..8}:
```

The expression i*1111$i=1..8, which makes use of the *sequence operator* $, is shorthand for the expression seq(i*1111,i=1..8)

```
> cycle:={}:
```

The variable cycle, initialized to the empty set, is to contain the first element of each repeating block (also called a *limit cycle*). The computation requires a double do-loop: the outer loop runs through all initial conditions x0, while the inner loop computes the corresponding orbit. The latter will be represented as a *set,* since we are only interested in detecting the beginning of periodicity.

```
> for x0 in ic do
> #----------- compute orbit starting at x0, until repetition
>     orbit:={x0}:
>     x:=F(x0,n):
>     while not member(x,orbit) do
>         orbit:= orbit union {x}:
>         x:=F(x,n)
```

```
>   od:
> #----------------------- store first element of cycle
>   cycle:=cycle union {x}
> od:
> cycle;
```

$$\{6174\}$$

We have shown that all orbits eventually reach the cycle beginning with 6174, which we know to have period 1. In the parlance of the theory of dynamical systems, we shall say that $\{6174\}$ is a *global attractor* for the system (9.3).

Exercises _____

Exercise 9.9. Let S be the set of all *finite* non-empty subsets of $\mathbb{Z}$. Let $s \in S$. Because s is finite and non-empty, it has a largest element M and a smallest element m. We consider the set s' obtained from s by replacing m by $m + 1$ and M by $M - m$. This is also finite and non-empty, and it is uniquely determined by s. So we have a function $f : s \mapsto s'$ of S into itself, given by

$$f : s \mapsto f(s) = ((s \setminus \{m\}) \setminus \{M\}) \cup \{m + 1\} \cup \{M - m\}. \qquad (9.4)$$

We construct recursive sequences in S, via f

$$s_0 = s \qquad\qquad s_{t+1} = f(s_t), \quad t \geq 0. \qquad (9.5)$$

For example, the initial condition $s_0 = \{-1, 2, 4\}$ leads to an eventually periodic sequence, with transient length 5 and period 3

$$\{-1, 2, 4\} \mapsto \{0, 2, 5\} \mapsto \{1, 2, 5\} \mapsto \{2, 4\} \mapsto \{2, 3\}$$
$$\mapsto \{1, 3\} \mapsto \{2\} \mapsto \{0, 3\} \mapsto \{1, 3\} \mapsto \cdots$$

(a) Implement the function f in (9.5) with Maple.

(b) Explore the behaviour of the sequence (9.5).

(c)* Prove that the sequence (9.5) is eventually periodic, for any choice of s_0.

Exercise 9.10. Let x be a positive integer. We construct a new integer $f(x)$ according to the following rule

$$f(x) = \begin{cases} 3x + 1 & \text{if } x \text{ is odd} \\ x/2 & \text{if } x \text{ is even}. \end{cases} \qquad (9.6)$$

Explore the behaviour of the associated recursive sequence. This is the celebrated '$3x + 1$' problem. The main conjecture — that all orbits eventually reach the cycle $(4, 2, 1)$ — is still unproved.

Exercise 9.11*. Prove that the map F_n of equation (9.3) is well-defined, in the sense that if x is an n-digit integer whose digits are not all the same, so is $F_n(x)$.

9.3 Conditional execution*

Conditional execution, represented by the `if`-structure, implements the process of decision-making in a programming language. Decisions are made based on the truth or falsehood of logical expressions. We have already met the `if`-structure in chapter 3, to construct characteristic functions.

In the simplest instance, the `if`-structure assumes the following form

```
> if logical expression then
>       statements
> fi;
```

If the value of the *logical expression* is *true*, the *statements* are executed. Otherwise, the control jumps to the statement following the `fi` statement.

The general form of the `if`-structure is the following:

```
> if logical expression 1 then
>       statements 1
> elif logical expression 2 then
>       statements 2
            ⋮
> else
>       default statements
> fi;
```

The *logical expression 1* is evaluated. If it is *true*, then *statements 1* are executed, and the structure is exited. Otherwise, *logical expression 2* is evaluated, and so on, sequentially for all `elif` statements (if any). Finally, if there is an `else` statement and all logical expressions in the structure are *false*, the *default statements* are executed. The logical expressions are evaluated automatically, without the need of using `evalb`.

Example 9.7. *Piecewise-defined functions.* These are functions whose value is given by distinct expressions in distinct regions of the domain. One such function is that converting marks (typically rational numbers between 0 and 100) into grades (typically elements of the set $\{A, B, C, D, E, F\}$), e.g.,

$$f(x) = \begin{cases} F & \text{if } 0 \le x < 40 \\ E & \text{if } 40 \le x < 50 \\ D & \text{if } 50 \le x < 60 \\ C & \text{if } 60 \le x < 70 \\ B & \text{if } 70 \le x < 80 \\ A & \text{if } 80 \le x \le 100. \end{cases} \tag{9.7}$$

We shall agree to return the NULL value when the argument x is out of range.

```
> f:=x->if x < 0 or x > 100 then
>          NULL
>       elif x < 40 then
>          F
>       elif x < 50 then
>          E
>       elif x < 60 then
>          D
>       elif x < 70 then
>          C
>       elif x < 80 then
>          B
>       else
>          A
> fi:
```

The first logical expression is *true* only if x is out of range. If this is not the case, then $x \geq 0$, necessarily, so that the single inequality $x < 40$ suffices to check the two inequalities in (9.7), and so on. If the control reaches the keyword else, then all previous logical expressions in the structure are *false,* whence $80 \leq x \leq 100$, and no further check is required for the grade A.

(An alternative way of constructing piecewise-defined functions is afforded by the standard library function piecewise.)

Example 9.8.   We develop an algorithm for finding a rational approximation to an irrational number lying in the unit interval $[0, 1]$. The algorithm is based on repeated construction of mediants (the mediant of two rationals was defined in (2.20)). We choose $\alpha = (\sqrt{5} - 1)/2$. (This number has a special status: it is the irrational between 0 and 1, which is the most difficult to approximate by rationals.) We begin by defining the function med for the mediant, and the small parameter small, which represents the desired accuracy of the result.

```
> med:=(l,r)->(numer(l)+numer(r))/(denom(l)+denom(r)):
> small:=1/1000:
> alpha:=evalf((sqrt(5)-1)/2):
```

The variable alpha has been assigned a *float* value, in order to allow the evaluation of the expression abs(m-alpha) >= small to Boolean (see below). The starting interval is the unit interval, with endpoints left and right. The iteration is carried out within a do loop controlled via the while option.

```
> left:=0:
```

```
> right:=1:
> m:=med(left,right):
> while abs(m-alpha) >= small do
>    if m < alpha then
>       left:= m
>    else
>       right:=m
>    fi:
>    m:=med(left,right)
> od:
> m;
```

$$\frac{21}{34}$$

We have implicitly assumed that the expression `abs(m-alpha) >= small` will eventually evaluate to *false,* thereby ensuring that the loop is not executed indefinitely (see exercises).

Exercises

Exercise 9.12. Consider the following irrational numbers in the unit interval ($e = 2.718\ldots$ is Napier's constant)

$$\frac{\sqrt{5}-1}{2}, \qquad \sqrt{2}-1, \qquad \pi - 3, \qquad e - 2.$$

By repeated iterations of the mediant method, determine rationals approximating these numbers within 10^{-6}. Compare the size of denominators.

Exercise 9.13*.   Prove that for any choice of α in the unit interval, the do-loop of the mediant algorithm does eventually terminate.

9.4   Procedures*

A function defined by means of the arrow operator `->` is a simple instance of a more general construct called a *procedure*, which we shall briefly describe in this section. We recall the syntax of the arrow operator

```
> name:=(arguments_sequence)->expression;
```
The following function adds 1 to each operand of an arbitrary object

```
> AddOneTo:=anything->map(x->x+1,anything):
```
The same function can be defined as a Maple *procedure* as follows

```
> AddOneTo:=proc(anything)
>    map(x->x+1,anything)
> end:
```
and then called as before

> AddOneTo({1,0,-22,b}), AddOneTo(a=b);

$$\{b+1, -21, 1, 2\}, a + 1 = b + 1$$

The range of valid data types for the argument **anything** is broad, but not quite arbitrary. For instance, the function call AddOneTo([{a}]); would produce an error termination, resulting from the attempt of adding an integer to a set.

The **proc** construct allows the user to define functions comprising more than one statement. A procedure would normally feature a control of the arguments' type, as well as the use of local variables. The syntax is the following:

> *name*:= proc(*arguments::type_sequence*);
>   [local *variables_sequence*;]
>   [option *options_sequence*;]
>   *statements*
> end;

The *arguments::type_sequence* is an expression sequence containing the arguments' names. Each name may (and indeed should) be followed by its required data type, separated by *two* colons. If more than one data type is acceptable, these should be listed as elements of a *set*.

Local variables are used for temporary storage. They are invisible from outside the procedure and unrelated to any variable with the same name that may occur elsewhere.

The meaning of the **option** sequence will be briefly illustrated below, in the context of recursive procedures.

Example 9.9.   The following procedure transforms the list of coefficients of a polynomial into a polynomial in the indeterminate x.

> Coeff2Poly:=proc(polylist::list, x::symbol)
> local i, degree:
> degree:=nops(polylist):
> 0:
> for i to degree do
>   % + polylist[i] * x^(degree - i)
> od
> end:

> Coeff2Poly([1,1,c^2+1,1], y);

$$y^3 + y^2 + (c^2 + 1)\, y + 1$$

Example 9.10.   The following procedure generates the list of primes lying between two integers a and b (a and b included).

> PrimesBetween:=proc(a::integer, b::integer)
> local L, p:

```
>
> L:=[]:
> p:=nextprime(a-1):
> while p <= b do
>    L:=[op(L),p]:
>    p:=nextprime(p)
> od:
> L
> end:
> PrimesBetween(1000,1100);
```

$$[1009, 1013, 1019, 1021, 1031, 1033, 1039, 1049, 1051, 1061, 1063,$$
$$1069, 1087, 1091, 1093, 1097]$$

The argument a must be an integer, being used by the function `nextprime`. This requirement is extended to the argument b for consistency. The two variables p and L are local. The initialization `p:=nextprime(a-1)` ensures that if a is prime, then it is included in the list. The output is stored in the variable L, which is initialized to the empty list. The reader is invited to compare the running time of this function with that of the simpler but slower construct developed in chapter 6, using `select`. (To compute run time, see `?time`.)

Example 9.11. We construct a function `freq` that computes the frequency of occurrence of the elements of an arbitrary list. For instance, in the list `[a,b,ba,b,ba,b,ab,b]`, the elements a and ab appear once, ba twice, and b four times.

```
> freq([a,b,ba,b,ba,b,ab,b]);
```

$$[a, 1], [b, 4], [ba, 2], [ab, 1]$$

The output is to be organized in the form of a sequence of two-elements lists, containing each value with its frequency.

```
> freq:=proc(L::list)
>    local l, x;
>    seq([x, nops(select((l,x)->evalb(l=x), L, x))], x={op(L)})
> end:
```

The expression `x={op(L)}` causes the control variable x in `seq` to run through all *distinct* elements of the list L, which is turned into a set to eliminate repetitions. The Boolean function `(l,x)->evalb(l=x)` is applied by the function `select` to all elements of L. Its second argument x is supplied as the third optional argument of `select`. The result of the selection is the list of those elements of L which are equal to x. Finally, the latter are counted by `nops`.

As an application, we consider the sequence of remainders of division of the first 1000 *odd* primes by 4. For instance, if we divide by 4 the primes

11 and 13, we obtain the remainders 3 and 1, respectively. We count their frequency.

```
> [seq(irem(ithprime(n),4),n=2..1001)]:
> freq(%);
```

$$[1, 495], [3, 505]$$

What have we learned from this experiment? First, the only possible remainders seem to be 1 and 3 (this is easy to explain: think about it). Second, the primes of the form $4n+1$ (remainder 1) and those of the form $4n+3$ (remainder 3) seem to be equally abundant (this is a celebrated theorem of number theory).

Recursive procedures

A procedure is said to be *recursive* if it calls itself.

As an application, we consider the problem of representing the element X_t of a first-order recursive sequence as an explicit function of t

$$X_0 = X \qquad X_t = f(X_{t-1}), \qquad t \geq 1. \tag{9.8}$$

Here f is a given function, and the initial value X is a given constant (an element of the domain of f). The element X_t is computed from X_0 by iterating the recursion (9.8) t times, which we have done earlier in this chapter by means of the do-construct. The use of recursive function call amounts to a literal translation into Maple of equation (9.8). It is computationally equivalent to a do-loop, but logically simpler.

We consider the recursive sequence of rational numbers

$$X_0 = 1 \qquad X_t = f(X_{t-1}) = \frac{X_{t-1} + 1}{X_{t-1} + 2}, \qquad t \geq 1. \tag{9.9}$$

We first define the function f

```
> f:=x->(x+1)/(x+2):
```

Then we write a procedure for the function $t \mapsto X_t$

```
> X:=proc(t::nonnegint)
> option remember:
> if t = 0 then
>    1
> else
>    f(X(t-1))
> fi
> end:
```

The argument t is a non-negative integer. The statement option remember causes Maple to remember the intermediate values of the recursion, avoiding re-computations. The rest of the procedure is a direct translation of equation (9.9).

Example 9.12. The following recursive procedure returns the *leaves* (innermost operands) of the tree representing a Maple expression x, in the form of an expression sequence.

```
> leaves:=proc(x)
> local y
> if (nops(x) = 1) then
>    x
> else
>    seq(leaves(y), y=x)
> fi
> end:
```

If x has more than one operand, we apply the function leaves recursively to all operands.

Exercises

Exercise 9.14*. Let $(x_1, y_1), \ldots, (x_4, y_4)$ be four points on the plane, with integer or rational coordinates. Write a procedure that returns *true* if these points are vertices of a square, and *false* otherwise.

Chapter 10

Vector spaces

This chapter contains some elements of linear algebra: the construction of a vector space (in the limited context of direct product of fields), the arithmetic of its elements (the *vectors*), and the study of a most important class of functions defined over vector spaces (represented by *matrices*).

10.1 Cartesian product of sets

We describe a general procedure for constructing new sets from old ones, which is analogous to the construction of a two-dimensional plane or a three-dimensional space, from a one-dimensional line.

Let T be a set. We denote by $T^2 = T \times T$ the set of *ordered pairs* of elements of T, called the *Cartesian product* of T with itself. Thus, T^2 is the collection of elements $v = (v_1, v_2)$ with $v_1, v_2 \in T$; the requirement that the pair be ordered means that $(v_1, v_2) \neq (v_2, v_1)$. Similarly, one defines $T^3 = T \times T \times T$, the set of ordered triples (v_1, v_2, v_3) of elements of T, and in general the set T^n of ordered n-tuples

$$T^n = \underbrace{T \times T \times \cdots \times T}_{n \text{ times}} = \{(v_1, v_2, \ldots, v_n)\} \qquad v_k \in T, \quad k = 1, \ldots, n.$$

The elements of T^n are called *vectors*.

Example 10.1. *Vectors.*

$$(1, -1, 3) \in \mathbb{Z}^3 \qquad\qquad (\tfrac{1}{3}, 1) \in \mathbb{Q}^2$$
$$(\sqrt{2} + i, 3i) \in \mathbb{C}^2 \qquad (x^2 + x + 1, 2x - 1, x, -x) \in \mathbb{Z}[x]^4.$$

If $T \subset U$, then $T^n \subset U^n$, for all positive n (why?). In particular, for every positive integer n we have the chain of inclusions

$$\mathbb{N}^n \subset \mathbb{Z}^n \subset \mathbb{Q}^n \subset \mathbb{R}^n \subset \mathbb{C}^n.$$

Example 10.2.   Let $T = \{0, 1\}$. Then T^3 has $2^3 = 8$ elements

$$T^3 = \{(0,0,0), (0,0,1), (0,1,0), (0,1,1), (1,0,0), (1,0,1), (1,1,0), (1,1,1)\}.$$

Example 10.3.   Let $I = [0, 1]$ be the unit interval

$$I = \{x \in \mathbb{R} \mid 0 \le x \le 1\}. \tag{10.1}$$

Then $I^2 \subset \mathbb{R}^2$ is a square of unit area, while $I^3 \subset \mathbb{R}^3$ is a cube of unit volume.

In the definition of Cartesian product of sets, there is no requirement that the sets involved be identical copies of the same set. Thus, $T \times S$ is defined as the set of ordered pairs $v = (t, s)$ with $t \in T$ and $s \in S$.

Example 10.4.   Let $T = \{1, 2, 3\}$, and $U = \{-1, -2\}$. Then

$$T \times U = \{(1,-1), (1,-2), (2,-1), (2,-2), (3,-1), (3,-2)\}.$$

Example 10.5.   Let T be a line and S a circle. Then $T \times S$ is an infinite cylinder. If I is the unit interval (10.1), then $I \times S$ is an annulus.

Example 10.6.   Let T be any set, and $S = \{1, 2, \ldots, n\}$. Then $T \times S$ can be thought of as the collection of n copies of T, indexed by the integers $1, 2, \ldots, n$.

10.2   Vector spaces

We introduce the notion of a vector space V, in the context of a Cartesian product of fields. This will suffice for our purposes. For a more general definition, consult the last chapter of this book.

Let F be a *field* (for instance, $F = \mathbb{Q}, \mathbb{R}, \mathbb{C}, \mathbb{Q}(x)$, etc.). Then F is equipped with the two arithmetical operations of sum and multiplication, which can be performed unrestrictedly along with their inverses (subtraction and division), excluding division by zero. These operations in F allow us to define certain operations in the Cartesian product $V = F^n$, which will make it into a *vector space*. In V, the elements of F will be called *scalars*.

We begin with the case $n = 2$. Let $V = F^2$, let $u = (u_1, u_2)$ and $v = (v_1, v_2)$ be elements of V, and let $\alpha \in F$. We define the sum $u + v$ and the scalar multiplication αv as follows

$$
\begin{aligned}
\textit{addition:} &\quad u + v = (u_1, u_2) + (v_1, v_2) = (u_1 + v_1, u_2 + v_2) \\
\textit{scalar multiplication:} &\quad \alpha u = \alpha \, (u_1, u_2) = (\alpha \, u_1, \alpha \, u_2).
\end{aligned}
$$

These operations satisfy certain properties, which derive directly from the arithmetic in a field. We list them below, leaving some of the proofs as exercises.

1. The sum is *commutative*

$$u + v = v + u.$$

Indeed,

$$
\begin{aligned}
u + v &= (u_1, u_2) + (v_1, v_2) = (u_1 + v_1, u_2 + v_2) = (v_1 + u_1, v_2 + u_2) \\
&= (v_1, v_2) + (u_1, u_2) = v + u.
\end{aligned}
$$

2. For all $u, v, w \in V$ we have the *associative law*:

$$(u + v) + w = u + (v + w)$$

which says that the order with which addition of vectors is performed does not matter.

3. The element $O = (0, 0)$ is such that for each v we have $v + O = O + v = v$:

$$
\begin{aligned}
v + O &= (v_1, v_2) + (0, 0) = (v_1 + 0, v_2 + 0) = (v_1, v_2) \\
&= (0 + v_1, 0 + v_2) = (0, 0) + (v_1, v_2) = O + v.
\end{aligned}
$$

4. The element $(-1)\, v$ is such that $v + (-1)\, v = O$:

$$
\begin{aligned}
v + (-1)v &= (v_1, v_2) + (-1)(v_1, v_2) = (v_1, v_2) + (-1v_1, -1v_2) \\
&= (v_1, v_2) + (-v_1, -v_2) = (v_1 - v_1, v_2 - v_2) = (0, 0) = O.
\end{aligned}
$$

5. The following distributive laws hold:

$$\alpha\,(u + v) = \alpha\,u + \alpha\,v, \qquad (\alpha + \beta)\,v = \alpha\,v + \beta\,v, \qquad (\alpha\beta)\,v = \alpha\,(\beta\,v).$$

6. There is an identity for scalar multiplication: $1\,v = v$.

The generalization of all the above definitions to the case of arbitrary positive n is straightforward. If $u = (u_1, \ldots, u_n)$ and $v = (v_1, \ldots, v_n)$, we have

$$u \pm v = (u_1 \pm v_1, \ldots, u_n \pm v_n) \qquad \alpha\,v = (\alpha v_1, \ldots, \alpha v_n).$$

A Cartesian product $V = F^n$ with the operations described above is an example of a *vector space* of dimension n over F. We remark that vector spaces are more general than Cartesian products of fields.

In a vector space, multiplication by a scalar combines a vector with a scalar, to yield another vector. There is no *multiplication* between two vectors that resembles multiplication between numbers. There is instead the *scalar product*, which is a binary operation turning a pair of vectors into a scalar. As before, we define it in the case in which V is a Cartesian product of fields.

Let $u = (u_1, \ldots, u_n)$ and $v = (v_1, \ldots, v_n)$ be two vectors in $V = F^n$. The scalar product of u and v is defined as

$$u \cdot v = u_1 v_1 + u_2 v_2 + \cdots + u_n v_n = \sum_{k=1}^{n} u_k\, v_k.$$

The scalar product satisfies the following properties, whose verification is left as an exercise.

1. $u \cdot v \in F$.
2. $u \cdot v = v \cdot u$.
3. $u \cdot (v + w) = u \cdot v + u \cdot w$.
4. $(\alpha u) \cdot v = \alpha(u \cdot v)$; $u \cdot (\alpha v) = \alpha(u \cdot v)$.
5. If $v = O = (0, 0, \ldots, 0)$, then $v \cdot v = 0$. Otherwise, $v \cdot v > 0$.

A distance in a vector space

The size of a vector $v = (v_1, \ldots, v_n)$ in V is called the *norm*, which is defined as

$$\|v\| = \sqrt{v_1^2 + v_2^2 + \cdots + v_n^2} = \sqrt{v \cdot v}. \tag{10.2}$$

As usual, we define the distance between two vectors as the norm of their difference: $\|v - u\|$. For $F = \mathbb{R}$, this distance becomes the ordinary Euclidean distance. Furthermore, for $n = 1$, the norm becomes the absolute value.

There are other ways of defining the norm of a vector, whence a distance in F^n. Specifically, for any positive real number p we define the p-norm of v as

$$\|v\|_p = \left(|v_1|^p + |v_2|^p + \cdots + |v_n|^p \right)^{1/p}. \tag{10.3}$$

Thus, the norm defined in (10.2) is the 2-norm ($p = 2$). The case $p = 1$ is also interesting. The 1-norm of a vector is the sum of the absolute value of the coordinates.

$$\|v\|_1 = |v_1| + |v_2| + \cdots + |v_n|.$$

Exercises

Exercise 10.1. Complete the proofs of the properties 1–6 of a vector space, first for $n = 2$, and then for arbitrary n.

10.3   Vectors with Maple

Arithmetic in vector spaces requires a specific set of Maple functions, which belong to the linear algebra library `linalg`. The easiest way to proceed is to load this library explicitly, using the intrinsic function `with`

```
> with(linalg):
```

(With the semicolon, all functions being loaded are listed.) We begin constructing an unspecified 2-dimensional vector $v = (v_1, v_2)$.

```
> v:=array(1..2);
```

$$v := array(1..2, [\,])$$

```
> v,whattype(v);
```

$$v, \; symbol$$

Unlike for other data types, the symbolic name of an array evaluates to itself (a symbol). Full evaluation requires a call to the intrinsic function evalm (evaluate to matrix)

```
> evalm(v);
```

$$\begin{bmatrix} v_1 & v_2 \end{bmatrix}$$

```
> whattype(%);
```

$$array$$

In computational parlance, we say that the array's name represents a *pointer* to the location in memory where the array information is stored. The function evalm is used to fetch that information. (A procedure's name has a similar status; see section 3.4.)

With the function op we can get some insight into the internal representation of an array in Maple

```
> op(evalm(v));
```

$$1..2, \begin{bmatrix} 1 = v_{[1]}, 2 = v_{[2]} \end{bmatrix}$$

Thus, an array consists of a range, which specifies the dimension, and a list, where each entry is assigned a value.

Next we construct a specific vector in $\mathbb{Z}^2$

```
> v:=array([1,-4]);
```

$$v := \begin{bmatrix} 1 & -4 \end{bmatrix}$$

The internal representation of v has now changed

```
> op(evalm(v));
```

$$1..2, [1 = 1, 2 = -4]$$

All components of an array may be accessed and assigned new values

```
> v[2];
```

$$-4$$

```
> v[2]:=-3:evalm(v);
```

$$\begin{bmatrix} 1 & -3 \end{bmatrix}$$

A vector can be multiplied by a scalar in a straightforward manner

```
> -2*v:%=evalm(%);
```

$$-2v = [-2 \quad 6]$$

Likewise, vectors can be easily summed and subtracted

```
> u:=array([0,1]):
> z:=evalm(u+v);
```

$$z := [1 \quad -2]$$

```
> whattype(z),whattype(evalm(z));
```

$$symbol, \ array$$

In the last example, the new array z was generated without resorting to the
array command, by simply assigning to the name z a datum of the *array*
type. By contrast, the assignment

```
> w:=u+v:
```

assigns to w the algebraic expression u+v. Evaluation of z and w gives the
same result

```
> evalm(z),evalm(w);
```

$$[1 \quad -2], [1 \quad -2]$$

but whereas z points to an array with a specified value, w points to the sum
of two variables which in turn point to arrays. Thus, if we change the value
of v

```
> v[1]:=99:
```

the value of w changes, but that of z does not

```
> evalm(z),evalm(w);
```

$$[1 \quad -2], [99 \quad -2]$$

The scalar product — also called *dot product* — is supported by the
linalg function dotprod

```
> dotprod(u,v);
```

$$-3$$

The p-norm of a vector (cf. equation (10.3)) is given by the function norm,
with p as a second argument

```
> norm(v,2);
```

$$\sqrt{10}$$

```
> w:=array(1..3):norm(w,4);
```

$$\left(|w_1|^4 + |w_2|^4 + |w_3|^4 \right)^{1/4}$$

Example 10.7.   Let $u = (-2, -3, 1) \in \mathbb{Z}^3$. Consider the following recursive sequence $v_0, v_1, \ldots$ of elements of $\mathbb{Z}^3$

$$v_0 = (1, 0, 0); \qquad v_{t+1} = f(v_t) = 2\,v_t + u, \quad t \geq 0.$$

We show that the first 10 elements of this sequence are orthogonal to the
vector $w = (0, 1, 3)$.

```
> u:=array([-2,-3,1]):
> w:=array([0,1,3]):
> v:=array([1,0,0]):
> f:=v->evalm(2*v+u):
> for t to 10 while dotprod(v,w) <> 0 do
>    v:=f(v)
> od:
> t;
```

$$11$$

As usual, all the information is buried in the function f that performs the
recursion. There is no need of displaying any intermediate computation.
Failure of orthogonality would have resulted in a smaller value of t (make
sure you understand this).

10.4 Matrices

A matrix over T is a rectangular array of elements of a given set T

$$A = \begin{pmatrix} 2 & 3 \\ -5 & 7 \end{pmatrix} \qquad B = \begin{pmatrix} c & -1 & c^2 - 3 \\ c^2 + c + 1 & 5c & -c \end{pmatrix}.$$

In the above example, A is a 2×2 matrix over $\mathbb{Z}$, and B is a 2×3 matrix over
$\mathbb{Z}[c]$ (the set of polynomials with integer coefficients in the indeterminate
c). An element a_{ij} of a matrix A is identified by two integers i and j,
specifying its position within the array: the row and the column number.
In the above example, we have $a_{21} = -5$, and $b_{13} = c^2 - 3$. To indicate
that a_{ij} are the elements of A, one writes $A = (a_{ij})$.

Matrices, like numbers, are algebraic objects, which combine according
to two binary operations: addition and multiplication. To define these
operations one must consider matrices over a fixed set T (usually a *ring* or
a *field*), and then impose constraints on the dimensions of such matrices.

The sum (subtraction) $A \pm B$ of two matrices A and B is performed
componentwise, so it can be defined only if the summands have the same
dimensions. Specifically, if $A = (a_{i,j})$ and $B = (b_{i,j})$ are $m \times n$ matrices,
and $C = (c_{i,j}) = A \pm B$, we define

$$c_{i,j} = a_{ij} \pm b_{ij} \qquad i = 1, \ldots, m, \qquad j = 1, \ldots n.$$

The sum of matrices behaves much like the sum of, say, the integers. In
particular, it enjoys the commutative and associative properties; i.e., $A + B = B + A$, and $(A + B) + C = A + (B + C)$.

The product $C = AB$ is defined when the column dimension of A is the same as the row dimension of B. In this case let n be such common dimension. The entry $c_{i,j}$ of C is defined as the scalar product of the ith row of A and the jth column of B

$$c_{ij} = a_{i1}b_{1j} + a_{i2}b_{2j} + \cdots + a_{in}b_{nj} = \sum_{k=1}^{n} a_{ik}b_{kj}.$$

If A and B are square matrices, then both AB and BA are defined, but in general we have that $AB \neq BA$. Failure of commutativity results in a very rich arithmetic.

Matrices may also represent certain *functions*, which assign to a vector v in a vector space V_1 the unique vector Av in another space V_2. (The multiplication between a matrix and a vector turns out to be a special case of multiplication between matrices.) The properties of matrices then become properties of these functions.

10.5   Matrices with Maple

In Maple, matrices are represented as many-dimensional arrays
> A:=array([[0,1],[3,-1]]);

$$A := \begin{bmatrix} 0 & 1 \\ 3 & -1 \end{bmatrix}$$

The argument of the function **array** is a list consisting of two lists, representing the first and second row of the array, respectively.
> A[2,1];

$$3$$

> -2*A:%=evalm(%);

$$-2A = \begin{bmatrix} 0 & -2 \\ -6 & 2 \end{bmatrix}$$

> B:=array([[2,0],[1,1]]);

$$B := \begin{bmatrix} 2 & 0 \\ 1 & 1 \end{bmatrix}$$

> A+B:%=evalm(%);

$$A + B = \begin{bmatrix} 2 & 1 \\ 4 & 0 \end{bmatrix}$$

Matrix multiplication is *non-commutative:* if A and B are n by n matrices, then $AB \neq BA$, in general. Maple supports the non-commutative multiplication operator **&***, which preserves the order of the operands.
> evalm(A&*B);

$$\begin{bmatrix} 1 & 1 \\ 5 & -1 \end{bmatrix}$$

```
> evalm(B&*A);
```

$$\begin{bmatrix} 0 & 2 \\ 3 & 0 \end{bmatrix}$$

```
> evalm(A*B);
```

Error, (in evalm/evaluate) use the &* operator for matrix/vector multiplication

Maple knows that the commutative multiplication operator * is ambiguous for matrices, and refuses to carry out the multiplication.

The following example illustrates further the differences between commutative and non-commutative multiplication, and clarifies Maple's strategy with expressions involving matrices.

```
> K:=array([[1,1],[1,0]]):
> L:=array([[2,1],[1,1]]):
```

The matrices K and L *commute*, that is, $KL = LK$.

```
> evalm(K&*L),evalm(L&*K);
```

$$\begin{bmatrix} 3 & 2 \\ 2 & 1 \end{bmatrix}, \begin{bmatrix} 3 & 2 \\ 2 & 1 \end{bmatrix}$$

Hence the *commutator* of K and L, given by $KL - LK$, is equal to the 2 by 2 null matrix

```
> evalm(K&*L-L&*K);
```

$$\begin{bmatrix} 0 & 0 \\ 0 & 0 \end{bmatrix}$$

However, if we instead use commutative multiplication, the result is quite different (and wrong!)

```
> evalm(K*L-L*K);
```

$$0$$

```
> whattype(%%),whattype(%);
```

array, integer

Using commutative multiplication, we have obtained the *integer* 0 rather than the 2 by 2 null matrix! Why? Also, why hasn't Maple complained — as it did before — when we used commutative multiplication with matrices? The reason is the following. Before accessing a matrix with `evalm`, Maple carries out some automatic algebraic simplifications, regarding K and L as symbols. (This could result in massive efficiency savings when dealing with large arrays.) Thus, Maple first transforms K*L into L*K, (or vice versa, depending on the context), which is legitimate because * is commutative.

Therefore, K*L-L*K evaluates to the integer 0, which is a valid argument for evalm, that treats it as a 1×1 matrix.

As for vectors, matrices can be generated directly by assigning to a variable the value of an evaluated matrix. In the next example, the value of the variable C is A, which points to an array, while the variable D points to a copy of the array A.

```
> C:=A:
> D:=evalm(A):
> A[1,1]:=22:
> evalm(C),evalm(D);
```

$$\begin{bmatrix} 22 & 1 \\ 3 & -1 \end{bmatrix}, \begin{bmatrix} 0 & 1 \\ 3 & -1 \end{bmatrix}$$

So changing A changes C but not D.

The library linalg supports a vast number of functions associated with matrix arithmetic

```
> M:=array([[a,b],[c,d]]);
```

$$M := \begin{bmatrix} a & b \\ c & d \end{bmatrix}$$

We begin with determinant and trace

```
> det(M),trace(M);
```

$$ad - bc, \ a + d$$

Next we construct the characteristic polynomial of M, in the indeterminate y

```
> charpoly(M,y);
```

$$y^2 - yd - ya + ad - bc$$

```
> collect(%,y);
```

$$y^2 + (\underbrace{-d - a}_{-\mathrm{Tr}(M)})\, y + \underbrace{ad - bc}_{\det(M)}$$

We list some linalg functions applied to a matrix with integer coefficients

```
> A:=([[0,1],[3,-1]]):
> evalm(A),trace(A),det(A),rank(A),transpose(A),
>         inverse(A),eigenvalues(A);
```

$$\begin{bmatrix} 0 & 1 \\ 3 & -1 \end{bmatrix}, -1, -3, 2, \begin{bmatrix} 0 & 3 \\ 1 & -1 \end{bmatrix}, \begin{bmatrix} \frac{1}{3} & \frac{1}{3} \\ 1 & 0 \end{bmatrix}, -\frac{1}{2} + \frac{1}{2}\sqrt{13}, -\frac{1}{2} - \frac{1}{2}\sqrt{13}$$

For more information about these functions, consult online documentation.

We consider some computation involving a 3 by 3 matrix over $\mathbb{Z}$. First we construct the matrix

```
> A:=array([[1,2,3],[4,5,6],[7,8,0]]);
```

$$A := \begin{bmatrix} 1 & 2 & 3 \\ 4 & 5 & 6 \\ 7 & 8 & 0 \end{bmatrix}$$

Next we verify that if $f(x)$ is the characteristic polynomial of A, then $f(A)$ is the null matrix (this is the *Hamilton-Cayley theorem*, valid for any square matrix)

```
> charpoly(A,z);
```

$$z^3 - 6\,z^2 - 72\,z - 27$$

```
> subs(z=A,%);
```

$$A^3 - 6\,A^2 - 72\,A - 27$$

```
> evalm(%);
```

$$\begin{bmatrix} 0 & 0 & 0 \\ 0 & 0 & 0 \\ 0 & 0 & 0 \end{bmatrix}$$

This can be done directly as follows

```
> evalm(charpoly(A,A));
```

$$\begin{bmatrix} 0 & 0 & 0 \\ 0 & 0 & 0 \\ 0 & 0 & 0 \end{bmatrix}$$

Maple multiplies any scalar by the identity matrix of the appropriate dimension before summing or subtracting it from a matrix

```
> evalm(%-2*x);
```

$$\begin{bmatrix} -2x & 0 & 0 \\ 0 & -2x & 0 \\ 0 & 0 & -2x \end{bmatrix}$$

We verify that the trace of A is the sum of its diagonal elements

```
> trace(A),add(A[k,k],k=1..3);
```

$$6,6$$

Summing all elements of A requires a double sum

```
> add(add(A[i,j],i=1..3),j=1..3);
```

36

Multiplying a matrix by a vector

Let the matrix A be as in the previous example.

```
> v:=array([1,0,-1]):
> evalm(A&*v);
```

$$[-2 \quad -2 \quad 7]$$

The function `evalm` interprets vectors as *column* vectors (even though it displays them as row vectors).

$$\begin{pmatrix} 1 & 2 & 3 \\ 4 & 5 & 6 \\ 7 & 8 & 0 \end{pmatrix} \begin{pmatrix} 1 \\ 0 \\ -1 \end{pmatrix} = \begin{pmatrix} -2 \\ -2 \\ 7 \end{pmatrix}.$$

This process can be iterated, to generate a recursive sequence of vectors

$$v_0 = (1, 0, -1) \qquad v_{t+1} = Av_t = f(v_t), \qquad t \geq 0.$$

The Maple implementation is structurally identical to that of any other recursive construction. All the information is buried in the definition of the function f

```
> f:=vct->evalm(A&*vct):
> to 4 do
>    v:=f(v)
> od;
```

$$[-2 \quad -2 \quad 7]$$
$$[15 \quad 24 \quad -30]$$
$$[27 \quad 0 \quad 297]$$
$$[864 \quad 1674 \quad -189]$$

Applying a function to all elements of an array

Let the matrix A be as in the previous example. We wish to square all elements of A, that is, to apply a function to all elements of a composite data type. This operation requires the use of the function `map`

```
> h:=y->y^2:
> map(h,A), map(z->z^2,A);
```

$$\begin{bmatrix} 1 & 4 & 9 \\ 16 & 25 & 36 \\ 49 & 64 & 0 \end{bmatrix}, \begin{bmatrix} 1 & 4 & 9 \\ 16 & 25 & 36 \\ 49 & 64 & 0 \end{bmatrix}$$

The following example illustrates the use of `map` with elements of $\mathbb{Z}[x]^2$.

```
> v:=array([x+1,x^2]):
> map(h,v);
```

$$[(x+1)^2 \quad x^4]$$

```
> map(expand,%);
```

$$[x^2 + 2x + 1 \quad x^4]$$

```
> x:=22:
> evalm(v);
```

$$[x + 1 \quad x^2]$$

```
> map(eval,v);
```

$$[23 \quad 484]$$

It should be noted that `map` acts on the *entries* of a matrix, even though the latter are not its *top-level operands* (see chapter 6). With `A` as above, we have

```
> op(evalm(A));
```

$$1..3, 1..3, [(1,1) = 1, (1,2) = 2, (1,3) = 3, (2,1) = 4, (2,2) = 5, \backslash$$
$$(2,3) = 6, (3,1) = 9, (3,2) = 8, (3,3) = 0]$$

```
> seq(whattype(x),x=[%]);
```

$$.., .., list$$

We see that the operands of a matrix are two ranges specifying the dimensions, and a list of equations defining the entries. An array is a special case of a more general construct called a `table`. See online documentation for more information.

Exercises

Exercise 10.2. Load the `linalg` package.

(*a*) Consider the following vectors in $\mathbb{Z}^3$

$$u = (88, -2, 33) \qquad v = (-7, 2^7, 1).$$

Compute their sum and their scalar product. Determine which one is the longest, using the 2-norm.

(*b*) Construct the following matrices

$$A = \begin{pmatrix} 1 & -2 \\ 2 & 0 \\ 3 & 1 \end{pmatrix} \qquad B = \begin{pmatrix} 1 & 1 & 1 \\ 4 & 6 & 2 \end{pmatrix}.$$

Then modify one entry of A as follows (without redefining the entire matrix!)

$$A = \begin{pmatrix} 1 & -2 \\ 2 & 0 \\ -5 & 1 \end{pmatrix} \qquad (10.4)$$

Then, using (10.4), compute BA, $A - B^T$, and $A^T + B$ (the superscript T denotes the transpose of a matrix).

Exercise 10.3.   Consider the recursive sequence of matrices

$$A_0 = \begin{pmatrix} 2 & 1 \\ 1 & 1 \end{pmatrix} \qquad A_{t+1} = A_t^2 - A_t - I_2, \qquad t \geq 0,$$

where I_2 is the 2×2 identity matrix.

(a) Compute A_5, using a do-loop. Do not display any intermediate output.

(b) Compute and display the elements of the sequence of integers $D_t = \text{trace}(A_t)$ for $t = 1, \ldots, 5$. No other output should be displayed.

Exercise 10.4.   The *commutator* $[A, B]$ of two square matrices A and B is defined as

$$[A, B] = AB - BA.$$

(Note that if A and B were numbers, $[A, B]$ would be zero.)

(a) Construct a user-defined function comm(A,B) whose value is $[A, B]$.

(b) Given the matrices

$$A = \begin{pmatrix} 0 & -1 \\ 1 & -1 \end{pmatrix} \qquad B = \begin{pmatrix} 0 & -1 \\ 1 & 0 \end{pmatrix}$$

we consider the sequence of matrices

$$A_n = A^n; \quad B_n = B^n; \quad C_n = [A, B]^n, \qquad n \geq 1.$$

By generating a sufficient number of elements of these sequences, convince yourself that they are all periodic and hence determine their period.

Exercise 10.5.   Let

$$M = \begin{pmatrix} 1 & 1 & 0 \\ 0 & 1 & 1 \\ 1 & 0 & 0 \end{pmatrix}. \tag{10.5}$$

(a) Construct a Maple function mu(M) for the function μ which subtracts 1 from all entries of a matrix M; e.g, if M is as above, then

$$\mu(M) = \begin{pmatrix} 0 & 0 & -1 \\ -1 & 0 & 0 \\ 0 & -1 & -1 \end{pmatrix}.$$

(b) Consider the recursive sequence of matrices

$$M_0 = M \qquad M_{t+1} = M_t + \mu(M_t), \qquad t \geq 0$$

Generate the first 5 elements of this sequence.

(c) By inspecting the result of the previous problem, *conjecture* the general form of M_t, valid for $t \geq 0$.

(*d*) Using Maple, prove by induction your conjecture of part (*c*). (This is a *computer-assisted proof.*)

(*e*) Let M be as in (10.5). Consider the recursive sequence of vectors of $\mathbb{Z}^3$

$$v_0 = (1,0,0) \qquad v_{k+1} = M\,v_k + v_0, \qquad k \geq 0.$$

Compute the smallest integer k for which the scalar product of v_k and v_0 exceeds 50.

Exercise 10.6. Let A and B be 3 by 3 matrices. Prove with Maple that

$$\mathrm{Det}(AB) = \mathrm{Det}(A)\,\mathrm{Det}(B).$$

[*Hint:* practice first with 2 by 2 matrices, until you are sure you can carry out the calculation without displaying any intermediate output.]

Exercise 10.7. Let

$$A(s) = \begin{pmatrix} -s & 0 & 0 & -24 \\ 1 & -s & 0 & 50 \\ 0 & 1 & -s & -35 \\ 0 & 0 & 1 & 10-s \end{pmatrix}.$$

Determine the values of s for which $A(s)$ is singular (non-invertible).

Chapter 11

Modular arithmetic*

Many arithmetical techniques involve operating with *modular* number systems, which are of importance in both theory and applications (data processing, coding theory, cryptography, etc.). Modular arithmetic deals with remainders of integer division. We have encountered these objects often, as values of the function `irem`.

This section contains an informal introduction to modular arithmetic with Maple. For further reading, see References [1, 2].

11.1   A modular system

We begin with arithmetic modulo the prime number 5. We consider the following five infinite sets of integers, called *congruence classes* modulo 5:

$$
\begin{aligned}
[0]_5 &= \{\ldots, -20, -15, -10, -5, \mathbf{0},\, 5,\, 10,\, 15,\, 20, \ldots\} \\
[1]_5 &= \{\ldots, -19, -14,\, -9, -4, \mathbf{1},\, 6,\, 11,\, 16,\, 21, \ldots\} \\
[2]_5 &= \{\ldots, -18, -13,\, -8, -3, \mathbf{2},\, 7,\, 12,\, 17,\, 22, \ldots\} \\
[3]_5 &= \{\ldots, -17, -12,\, -7, -2, \mathbf{3},\, 8,\, 13,\, 18,\, 23, \ldots\} \\
[4]_5 &= \{\ldots, -16, -11,\, -6, -1,\, 4,\, 9,\, 14,\, 19,\, 24, \ldots\}
\end{aligned}
\tag{11.1}
$$

By inspecting (11.1) we see that the first set $[0]_5$ is constructed starting from the integer 0, and then adding to it all multiples of 5. Thus,

$$[0]_5 = \{\ldots, 0-3\cdot5,\ 0-2\cdot5,\ 0-1\cdot5,\ 0+0\cdot5,\ 0+1\cdot5,\ 0+2\cdot5,\ 0+3\cdot5,\ \ldots\}.$$

Similarly

$$[1]_5 = \{\ldots, 1-3\cdot5,\ 1-2\cdot5,\ 1-1\cdot5,\ 1+0\cdot5,\ 1+1\cdot5,\ 1+2\cdot5,\ 1+3\cdot5,\ \ldots\},$$

etc., and in general, for $k \in \mathbb{Z}$ we have

$$[k]_5 = \{\ldots, k-3\cdot5,\ k-2\cdot5,\ k-1\cdot5,\ k+0\cdot5,\ k+1\cdot5,\ k+2\cdot5,\ k+3\cdot5,\ \ldots\}.$$

In the last equation, k is not restricted to the first five natural numbers. In particular, the same congruence class may be represented by different values of k. So, for instance, $[1]_5$, $[6]_5$, and $[-14]_5$ are the same class, as easily verified.

From the above considerations, it follows that every integer in $\mathbb{Z}$ appears in precisely one congruence class. Therefore, the *union* of these sets is the whole of $\mathbb{Z}$

$$[0]_5 \cup [1]_5 \cup [2]_5 \cup [3]_5 \cup [4]_5 = \mathbb{Z},$$

and they are pairwise disjoint — they have *empty intersection*

$$[0]_5 \cap [1]_5 = \emptyset, \qquad [0]_5 \cap [2]_5 = \emptyset, \qquad \text{etc.}$$

(We say that the 5 sets $[0]_5$, $[1]_5$, ... ,$[4]_5$ form a *partition* of $\mathbb{Z}$.)

We define the following *finite* set

$$\mathbb{F}_5 = \{[0]_5, [1]_5, [2]_5, [3]_5, [4]_5\}.$$

Thus, $\mathbb{F}_5$ consists of five elements, each of which is an *infinite* set.

If two integers a and b belong to the same element of $\mathbb{F}_5$, we say that 'a is congruent to b modulo 5', and write

$$a \equiv b \,(\text{mod } 5).$$

This expression is called a *congruence* (whence the name *congruence classes*). Thus,

$$5 \equiv 0 \,(\text{mod } 5) \qquad 6 \equiv 1 \,(\text{mod } 5) \qquad -2 \equiv 3 \,(\text{mod } 5).$$

It is plain that two integers x and y belong to the same congruence class precisely when they differ by a multiple of 5, that is, when $x - y$ gives remainder zero when divided by 5. Thus, -19 and 21 belong to the same class in $\mathbb{F}_5$, because their difference $-19 - 21 = -40 = (-8) \cdot 5$ is a multiple of 5. So we have $-19 - 21 \equiv 0 \,(\text{mod } 5)$, and since $-19 \equiv 1 \,(\text{mod } 5)$, we have $[-19]_5 = [1]_5$. Remainders of integer division are computed in Maple with the intrinsic function `irem` (see section 2.4)

```
> irem(-19-21,5);
```

$$0$$

Example 11.1. Some congruences.

$$2 + 3 \equiv 0 \,(\text{mod } 5) \qquad 2 - 4 \equiv 3 \,(\text{mod } 5) \qquad 2 \cdot 3 \equiv 1 \,(\text{mod } 5).$$

To understand the middle congruence, note that $5 \equiv 0 \,(\text{mod } 5)$, whence

$$-4 \equiv 0 - 4 \equiv 5 - 4 \equiv 1 \,(\text{mod } 5)$$

so $2 - 4 \equiv 2 + 1 \equiv 3 \,(\text{mod } 5)$.

Determining membership to a class is somewhat delicate. Membership is established by checking the remainder upon division by 5. However, according to the definition of remainder given in (2.11), two members of a class will produce the same remainder only if they agree in sign. This definition is consistent with the behaviour of `irem`, whose value can be positive as well as negative. Thus, 8, 13, and -7 belong to the same class $[3]_5$, yet their respective remainders are not the same

> `irem(8,5),irem(13,5),irem(-7,5);`

$$3, 3, -2$$

(There is no contradiction here, since $[3]_5 = [-2]_5$.)

This problem is solved if we agree that we shall always use a non-negative remainder. Maple has a specific binary operator for modular arithmetic, called `mod`, which returns the *least non-negative remainder* of an integer division

> `8 mod 5,13 mod 5,-7 mod 5;`

$$3, 3, 3$$

In the rest of this chapter, we shall always use `mod`.

All the above constructions and considerations can be generalized. We replace 5 with an arbitrary *prime* number p, and define

$$\mathbb{F}_p = \{[0]_p, [1]_p, \ldots, [p-1]_p\},$$

where the elements of $\mathbb{F}_p$ are infinite sets defined for every integer k as follows

$$[k]_p = \{\ldots, k-3\cdot p, k-2\cdot p, k-1\cdot p, k+0\cdot p, k+1\cdot p, k+2\cdot p, k+3\cdot p, \ldots\}.$$

Again, the integer k is not restricted to the first p natural numbers, so that different values of k may correspond to the same class

$$[k]_p = [k+sp]_p \qquad s \in \mathbb{Z}.$$

The requirement that p be prime is essential, as we shall see.

Example 11.2. 2^{21} and -10011 belong to the same congruence class $[89]_{101}$ of $\mathbb{F}_{101}$.

> `2^21 mod 101, -10011 mod 101;`

$$89, 89$$

We construct a function to test the congruence $x \equiv y \pmod{m}$:

> `AreCongruent:=(x::integer,y::integer,m::posint)`
>   `-> evalb(x = y mod m):`
> `AreCongruent(2^21,-10011,101);`

$$true$$

Note that the modulus m has data type `posint`, a positive integer.

Exercises

Exercise 11.1. Construct the sets $\mathbb{F}_2$, $\mathbb{F}_3$, and $\mathbb{F}_7$, as in (11.1).

Exercise 11.2. Verify the following congruences

$$-6 \equiv 5030497 \,(\mathrm{mod}\ 503); \quad -2^{10} + 101^3 - 29274 \equiv 0 \,(\mathrm{mod}\ 1000003).$$

Exercise 11.3. Determine the largest integer less than a million which belongs to the class $[777]_{1009}$ in $\mathbb{F}_{1009}$.

Exercise 11.4. Prove that the union of the elements of $\mathbb{F}_p$ is $\mathbb{Z}$, and that they are pairwise disjoint (they form a *partition* of $\mathbb{Z}$.)

11.2   Arithmetic of equivalence classes

We are ready to define the four arithmetical operations for congruence classes modulo a prime number. Addition, subtraction, and multiplication derive naturally from the corresponding integer operations. More surprising is the fact that *division* by a nonzero congruence class is also defined, in spite of the fact that the division of two integers does not yield an integer, in general. As a result, arithmetic modulo a prime number will enjoy many properties of rational arithmetic.

Addition, subtraction, and multiplication

Addition, subtraction, and multiplication can be performed without restrictions in $\mathbb{Z}$. Using this property, we now define the corresponding operations between congruence classes.

   To sum two congruence classes $[a]_p$ and $[b]_p$, we choose an *arbitrary* element from each, and then compute their sum, which is an integer s. Since the classes constituting $\mathbb{F}_p$ are disjoint, s must belong to precisely one class, and the latter is defined to be the sum of the two classes: $[a]_p + [b]_p = [s]_p$. It remains to be shown that this operation is well-defined, that is, that the result does not depend on the particular choice of a and b within the respective classes. Thus, let a' and b' be two other elements of $[a]_p$ and $[b]_p$, respectively. Then for some integers k and l we have $a' = a + pk$ and $b' = b + pl$, so that $a' + b'$ differs from $a + b$ by a multiple of p. But then $a' + b'$ and $a + b$ belong to the same class, by definition.

   The same procedure is used to define subtraction and multiplication. Thus, in $\mathbb{F}_5$ we have $[4]_5 + [3]_5 = [2]_5$. Indeed, choosing $-6 \in [4]_5$ and $18 \in [3]_5$, we find that $-6 + 18 = 12 \in [2]_5$.

```
> -6+18 mod 5;
```

$$2$$

   The integer 0 is the *identity element* of addition in $\mathbb{Z}$: $a + 0 = a$, for any integer a. It then follows that the class $[0]_p$ is the identity class of modular addition, because we can always choose 0 as a representative element of $[0]_p$.

By the same token, the integer 1 is the identity element of multiplication in $\mathbb{Z}$: $a \cdot 1 = a$, for any integer a, so that $[1]_p$ is the identity class of modular multiplication.

Division

Division cannot be performed unrestrictedly in $\mathbb{Z}$, even if we exclude division by zero. We approach this problem by attempting to construct the reciprocal $[1]_p/[b]_p$ of every nonzero element $[b]_p \in \mathbb{F}_p$. The quantity $[1]_p/[b]_p$ will be 'that thing which, when multiplied by $[b]_p$, gives $[1]_p$', called the *modular inverse* of $[b]_p$

$$\frac{[1]_p}{[b]_p} \cdot [b]_p = [1]_p. \tag{11.2}$$

In order to determine $[1]_p/[b]_p$, we must solve the congruence $x \cdot b \equiv 1 \,(\mathrm{mod}\ p)$. Then $1/b \equiv x \,(\mathrm{mod}\ p)$. We do this for all nonzero elements of $\mathbb{F}_5$.

$$
\begin{array}{lll}
x \cdot 1 \equiv 1 \,(\mathrm{mod}\ 5) & \implies \quad x \equiv 1 \,(\mathrm{mod}\ 5) & \implies \quad [1]_5/[2]_5 = [1]_5 \\
x \cdot 2 \equiv 1 \,(\mathrm{mod}\ 5) & \implies \quad x \equiv 3 \,(\mathrm{mod}\ 5) & \implies \quad [1]_5/[2]_5 = [3]_5 \\
x \cdot 3 \equiv 1 \,(\mathrm{mod}\ 5) & \implies \quad x \equiv 2 \,(\mathrm{mod}\ 5) & \implies \quad [1]_5/[3]_5 = [2]_5 \\
x \cdot 4 \equiv 1 \,(\mathrm{mod}\ 5) & \implies \quad x \equiv 4 \,(\mathrm{mod}\ 5) & \implies \quad [1]_5/[4]_5 = [4]_5.
\end{array}
$$

Maple knows this

```
> seq(1/k mod 5, k=1..4);
```

$$1, 3, 2, 4$$

There is another way of looking at the construction of the modular inverse. We know we cannot divide 1 by 2 in $\mathbb{Z}$. Does this mean that we cannot divide $[1]_5$ by $[2]_5$ in $\mathbb{F}_5$? The procedure we have used to define addition, subtraction, and multiplication made use of the corresponding operations in $\mathbb{Z}$ performed on *arbitrary* elements of the operand classes. So if we choose 1 from $[1]_5$ and 2 from $[2]_5$, we are stuck, because $1/2$ is not in $\mathbb{Z}$, and therefore it cannot be found in any class. However, if we instead choose 6 as the representative element of $[1]_5$, then we can perform the integer division: $6/2 = 3 \in [3]_5$. We obtain the same result by selecting, say, $21 \in [1]_5$ and $-3 \in [2]_5$, for $21/(-3) = -7 \in [3]_5$. So it would seem that division in $\mathbb{F}_5$ can be derived consistently from division in $\mathbb{Z}$, provided we choose *suitable* (rather than *arbitrary*) representative elements from the relevant classes, that is, elements for which such division yields an integer result.

It turns out that the above procedure works in general, due to the following

Theorem 10 *Let p be a prime number. Then, if $b \not\equiv 0 \,(\mathrm{mod}\ p)$, the congruence $x \cdot b \equiv 1 \,(\mathrm{mod}\ p)$ admits a unique solution modulo p.*

This theorem guarantees that for each class $[b]_p \neq [0]_p$ we can find a unique class $1/[b]_p$, satisfying (11.2). But then for each $[a]_p$ and nonzero $[b]_p$, the quantity $[a]_p/[b]_p$ is uniquely defined, since

$$\frac{[a]_p}{[b]_p} = [a]_p \cdot \frac{[1]_p}{[b]_p}.$$

The construction of the inverse of b rests on *Euclid's algorithm* for the greatest common divisor of two integers, according to which if b and p are coprime, we can find integers x and y such that $x \cdot b + y \cdot p = 1$, that is, $x \cdot b \equiv 1 \,(\mathrm{mod}\ p)$.

The existence of a multiplicative inverse ensures that for every prime p, the set $\mathbb{F}_p$ is a *finite field*. In other words, in $\mathbb{F}_p$, like in $\mathbb{Q}$, $\mathbb{R}$, $\mathbb{C}$, $\mathbb{Q}(x)$, etc., we can add, subtract, multiply, and divide, except division by zero. Thus, for instance, any fraction with denominator coprime to p makes sense in $\mathbb{F}_p$

> 7/5-41/4 mod 101;

77

It is important to note that the condition that p be prime is necessary for the above construction. For instance, the inverse of a nonzero element does not necessarily exist in the arithmetic modulo 6.

> 1/5 mod 6;

5

> 1/4 mod 6;

Error, the modular inverse does not exist

(Indeed, 4 is even, and so is any of its multiples, so that no multiple of 4 can be congruent to 1 modulo 6.) Modular arithmetic for a composite modulus is very rich and interesting, but we shall not consider it here.

11.3   Some arithmetical constructions in $\mathbb{F}_p$

In this section, we describe some standard constructions in a finite field $\mathbb{F}_p$: the squares and higher powers, exponentiation and the logarithm, polynomial arithmetic. (When no confusion can arise, we shall represent the congruence class $[a]_p$ via the unique integer $r \equiv a \,(\mathrm{mod}\ p)$, which lies in the range $0, \ldots, p-1$.)

Squares and higher powers

The function $f := x \mapsto x^2$ maps the integers into the *squares*

$$f(\mathbb{Z}) = \{0, 1, 4, 9, 16, 25, 36, \ldots\}$$

and every nonzero square is the image under f of two distinct elements of $\mathbb{Z}$.

The squares in $\mathbb{Z}$ form a 'thin set', in the sense that the probability that an integer taken at random is a square tends to zero as the integer becomes large. For instance, 40% of the integers less than 10 are squares, 10% of the integers less than 100 are squares, and only some 3% of the integers less than 1000 are squares.

The squares are also defined in $\mathbb{F}_p$, where the situation is quite different. We begin with the case $p = 7$.

```
> F7:={$0..6};
```

$$\{0, 1, 2, 3, 4, 5, 6\}$$

To square the elements of $\mathbb{F}_7$, we make use of the operator mod

```
> f:=x->x^2 mod 7:
```

and to apply the function f to every element of $\mathbb{F}_7$ we use the standard library function map

```
> map(f,F7);
```

$$\{0, 1, 2, 4\}$$

This calculation shows that half of the 6 nonzero elements in $\mathbb{F}_7$ are squares. Specifically, modulo 7, we have

$$1^2 \equiv 6^2 \equiv 1, \qquad 2^2 \equiv 5^2 \equiv 4, \qquad 3^2 \equiv 4^2 = 2.$$

Next we count the squares in $\mathbb{F}_{101}$

```
> {$0..100}:
> map(x->x^2 mod 101,%):
> nops(%)-1;
```

$$50$$

Thus, also in $\mathbb{F}_{101}$, half of the nonzero elements are squares. This result is true in general.

Theorem 11 *Let p be an odd prime. Then half of the nonzero elements of $\mathbb{F}_p$ are squares.*

The Maple number theory library numtheory features — among other things — a rich set of functions for modular arithmetic. This library is loaded with the command with

```
> with(numtheory):
```

(We have already seen this command in section 10.3.) Now we may freely use all numtheory functions; for instance, to ascertain whether or not a is a square modulo p, we use the function legendre(a,p), which returns $+1$ if a is a square in $\mathbb{F}_p$, -1 if a is not a square, and 0 if a is divisible by p.

This function is (essentially) the characteristic function of the squares in
$\mathbb{F}_p$, which is known as the *Legendre symbol*.

```
> legendre(7,47);
```

$$1$$

The above result shows that 7 is a square modulo 47. To compute its least
non-negative square root, we use the function `msqrt`

```
> msqrt(7,47);
```

$$17$$

(The other value of $\sqrt{7}$ is then $-17 \equiv 47 - 17 \equiv 30 \,(\mathrm{mod}\ 47)$.)

If b is an integer greater than 1, then the function $f_b := n \mapsto b^n$ will
map the elements of $\mathbb{F}_p$ into the bth powers

Example 11.3. Find all the squares in $\mathbb{F}_{29}$ which are not fourth powers.

```
> F29:={$0..28}:
> f:=(x,b,p)->x^b mod p:
> map(f,F29,2,29):
> map(f,F29,4,29):
> %% minus %;
```

$$\{4,\ 5,\ 6,\ 9,\ 13,\ 22,\ 28\}$$

Exponentiation and logarithm

We consider the process of exponentiation in $\mathbb{F}_p$: $x \mapsto b^x \,(\mathrm{mod}\ p)$, where
b is a fixed integer. As an example, we consider the case $p = 7$ and $b = 2$.
We find

$$2^0 \equiv 1 \,(\mathrm{mod}\ 7), \quad 2^1 \equiv 2 \,(\mathrm{mod}\ 7), \quad 2^2 \equiv 4 \,(\mathrm{mod}\ 7), \quad 2^3 \equiv 8 \equiv 1 \,(\mathrm{mod}\ 7)$$

whence $2^4 \equiv 2^1 \,(\mathrm{mod}\ 7)$, etc. Thus, the powers of 2 in $\mathbb{F}_7$ are 1, 2, and 4;
that is, the exponential sequence $n \mapsto [2^n]_7$ is *periodic*, with period 3.

In general, if a is not divisible by p, we call the (multiplicative) *order*
of a modulo p, to be the period of the sequence $n \mapsto [a^n]_p$. Since $[1] = [a^0]$
belongs to the repeating part, the order of a modulo p can also be defined as
the smallest positive integer n for which $a^n \equiv 1 \,(\mathrm{mod}\ p)$. The corresponding
Maple function is `order(a,p)`

```
> order(2,7)
```

$$3$$

However, the case $b = 3$ presents us with an interesting phenomenon. We
find, modulo 7

$$3^0 \equiv 1, \qquad 3^1 \equiv 3, \qquad 3^2 \equiv 2, \qquad 3^4 \equiv 4, \qquad 3^5 \equiv 5, \qquad 3^6 \equiv 1.$$

Thus, *all* nonzero elements of $\mathbb{F}_7$ can be expressed as some power of 3. It turns out that this phenomenon is generic: in every finite field there exists at least one element g whose powers generate all nonzero field elements. Such g is called a *primitive root* modulo p. Alternatively, a primitive root may be defined as an integer g whose order is equal to $p - 1$.

Example 11.4. The integer 2 is a primitive root modulo 11.

```
> seq(2^k mod 11, k=1..10);
```

$$2, 4, 8, 5, 10, 9, 7, 3, 6, 1$$

and 3 is a primitive root modulo 199

```
> for n while((3^n mod 199) > 1) do od:
> n;
```

$$198$$

The function `primroot(p)` of the `numtheory` library returns the smallest positive primitive root modulo a prime p

```
> primroot(9973);
```

$$11$$

Let us double-check

```
> seq(order(a,9973),a=1..11);
```

$$1, 3324, 831, 1662, 3324, 3324, 3324, 1108, 831, 554, 9972$$

It can be shown that the order of an element is always a divisor of $p - 1$

```
> map(x->9972 mod x,{%});
```

$$\{0\}$$

The existence of a primitive root g modulo p implies that the congruence

$$g^y \equiv a \pmod{p}$$

can always be solved for y, provided that a is nonzero. By analogy with the complex case, we shall then call x the *discrete logarithm* of a to the base g, modulo p. The function `mlog(a,g,p)` computes such logarithm

```
> mlog(500,11,9973);
```

$$2085$$

```
> 11^2085 mod 9973;
```

$$500$$

Exponentiation in modular arithmetic is a well-known *trapdoor function,* easy to do but very difficult to undo. In other words, the discrete logarithm is a lot more time-consuming than its inverse, the exponentiation. We have already encountered this phenomenon in regard to prime factorization (see section 2.6).

Polynomials

Since $\mathbb{F}_p$ is a field, defining polynomials with coefficient in $\mathbb{F}_p$ presents no conceptual difficulty: one represents the coefficients as integers (or rationals with denominator coprime to p). The operator mod will then reduce them to the range $0, \ldots, p-1$.

```
> 40*x^2-x/3-19 mod 13;
```

$$x^2 + 4x + 7$$

For polynomial divisibility, Maple provides the inert functions Factor, Quo, and Rem, to be used in conjunction with mod.

Let us consider the polynomial $g(x) = x^2 + 1$ in $\mathbb{F}_5$. Since $5 \equiv 0 \,(\mathrm{mod}\ 5)$ and $6 \equiv 1 \,(\mathrm{mod}\ 5)$, we have

$$g(x) = (x+2)\,(x+3) = x^2 + 5x + 6 \equiv x^2 + 1 \,(\mathrm{mod}\ 5).$$

We see that $g(x)$ factors in $\mathbb{F}_5$ as $(x+2)(x+3)$, much in the same way as it factors in $\mathbb{C}$ as $(x+i)(x-i)$, where $i = \sqrt{-1}$ is the imaginary unit. Therefore, $-2 \equiv 3$ and $-3 \equiv 2$ are the two square roots of -1 in $\mathbb{F}_5$.

```
> Fg:=p->Factor(x^2+1) mod p:
> Fg(5);
```

$$(x+2)\,(x+3)$$

Writing it all out in detail, we have

$$g(x) = [1]_5 x^2 + [1]_5 = (x - [2]_5)\,(x - [3]_5), \qquad g([2]_5) = g([3]_5) = [0]_5,$$

that is, g(x) has two distinct *roots* in $\mathbb{F}_5$

Let us look into the factorization of $x^2 + 1$ in $\mathbb{F}_p$, for other values of p

```
> seq(Fg(ithprime(k)),k=1..6);
```

$$(x+1)^2,\ x^2 + 1,\ (x+3)\,(x+2),\ x^2 + 1, x^2 + 1,\ (x+8)\,(x+5)$$

```
> map(whattype,[%]);
```

$$[\hat{\ },\ +,\ *,\ +,\ +,\ *]$$

Three types of factorizations occur, corresponding to $x^2 + 1$ having one double root (data type ^), two distinct roots (data type *), or no roots at all (data type +) modulo p. It turns out that these are the only possibilities. We explore this phenomenon for a large set of primes

```
> data:=map(whattype,[seq(Fg(ithprime(k)),k=1..1000)]):
```

Now the variable data contains the data type of the factored polynomials, which in this case is preferable to storing the polynomials themselves. We count the occurrence of each of the three types, using the function select in conjunction with nops

```
> seq(nops(select((x,y)->evalb(x=y),data,y)),y=['^','*','+']);
```

$$1, 495, 504$$

Thus, the double root occurs only once (for the prime $p = 2$), while the other two cases occur with roughly equal probability.

Exercises _____

Exercise 11.5. Determine the cubes in $\mathbb{F}_{101}$.

Exercise 11.6. Show that all elements of $\mathbb{F}_{29}$ have at least one cube root.

Exercise 11.7. Determine all primitive roots modulo 101.

Exercise 11.8. Convince yourself that for every positive integer k the polynomial $x^2 + 1$ has two distinct roots modulo 5^k. That is, there exist integers a_k and b_k such that $x^2 + 1 \equiv (x - a_k)(x - b_k) \,(\text{mod } 5^k)$. Represent a_k and b_k to the base p: what do you observe as k becomes large?

Exercise 11.9. By considering the factorization of the polynomial $x^3 - 2$ in $\mathbb{F}_p$ for a large set of primes p, provide evidence that the probability that 2 has three cube roots modulo p is $1/6$.

Chapter 12

Some abstract structures*

In this chapter, we present the axiomatic definition of some abstract structures that have been exemplified throughout the course.

12.1   The axioms of arithmetic

We list the fundamental properties of the integers (called *axioms*) from which most other properties can be derived. Let x, y, and z be integers. Then

I	$x + y$ and $x \cdot y$ are in $\mathbb{Z}$.	closure
II	$x + y = y + x$; $x \cdot y = y \cdot x$.	commutativity
III	$(x + y) + z = x + (y + z)$;	associativity
	$(x \cdot y) \cdot z = x \cdot (y \cdot z)$.	
IV	$x + 0 = x$; $x \cdot 1 = x$.	identity elements
V	$x \cdot (y + z) = x \cdot y + x \cdot z$.	distributivity
VI	For each $x \in \mathbb{Z}$ there exists $-x \in \mathbb{Z}$	inverse of addition
	such that $x + (-x) = 0$.	
VII	If $x \cdot y = x \cdot z$ and $x \neq 0$ then $y = z$	cancellation

Note that the integers 0 and 1 enjoy a special status, in that addition by 0 and multiplication by 1 do not alter the elements of $\mathbb{Z}$. The properties *IV* should indeed be taken as the *definition* of the identity elements of addition and multiplication, respectively. Any set satisfying these axioms (with 0 and 1 possibly replaced by two elements of the set) will have many properties in common with $\mathbb{Z}$. Sets of this type are called *integral domains:* they are the integers' closest relatives.

12.2   Metric spaces

A metric space is a set where we can measure the distance between any two elements. Specifically, a metric space is a set X equipped with a function $\rho(x, y)$ defined for all $x, y \in X$ which has the properties:

I	$\rho(x, y)$ is real and non-negative.	
II	$\rho(x, y) = 0$ if and only if $x = y$.	
III	$\rho(x, y) = \rho(y, x)$.	symmetry
IV	$\rho(x, z) \leq \rho(x, y) + \rho(y, z)$.	triangle inequality

Example 12.1.   The set $\mathbb{C}$ with distance $\rho(x, y) = |x - y|$ is a metric space. The subsets $\mathbb{N}$, $\mathbb{Z}$, $\mathbb{Q}$, and $\mathbb{R}$ of $\mathbb{C}$ inherit from $\mathbb{C}$ the distance ρ, whence the structure of metric spaces.

Example 12.2.   Let $\Sigma_2(n)$ be the set of binary sequences of n elements: $x = (x_1, x_2, \ldots, x_n)$, $x_k \in \{0, 1\}$. We define

$$\rho(x, y) = \sum_{k=1}^{n} |x_k - y_k|.$$

Then $\rho(x, y)$ is an integer between 0 and n. Because $\rho(x, y)$ is a finite sum of non-negative terms, it is zero only when all summands are zero, that is, when $x = y$. Properties III and IV follow from the corresponding properties of the function $|x - y|$. Thus, $\Sigma_2(n)$ with the distance ρ is a metric space.

12.3   Rings and fields

A set R is called a *ring* if two binary operations are defined in R, called sum ($+$) and multiplication ($\cdot$), satisfying the following conditions for all elements of R

I	$a + y = y + a$.	commutativity of sum
II	$(a + y) + z = a + (y + z)$;	associativity
	$(a \cdot y) \cdot z = a \cdot (y \cdot z)$.	
III	$a \cdot (y + z) = a \cdot y + a \cdot z$;	distributivity
	$(a + y) \cdot z = a \cdot z + y \cdot z$.	
IV	For all $a, b \in R$ there exists $x \in R$	injectivity of sum
	such that $a + x = b$.	

Letting $b = a$ in IV, we obtain the identity element $x = 0$ of addition. Letting $b = 0$, we obtain the additive inverse $x = -a$ of a.

Note that multiplication need not be commutative. If the commutative law holds for multiplication, i.e., if $a \cdot b = b \cdot a$, the ring is said to be *commutative*. If an element 1 exists such that $1 \cdot a = a \cdot 1 = a$ for all a, then R is said to be a *ring with unity*.

Example 12.3. The set of even integers

$$2\mathbb{Z} = \{\ldots, -4, -2, 0, 2, 4, \ldots\}$$

is a commutative ring without unity.

Example 12.4. The subset of $\mathbb{C}$ constituted by all complex numbers with integer real and imaginary parts forms a ring, called the ring of *Gaussian integers*, denoted by $\mathbb{Q}[i]$

$$\mathbb{Q}[i] = \{x + iy \mid x, y \in \mathbb{Z}, i = \sqrt{-1}\}.$$

Replacing the integer -1 under the square root with an integer D not a square, we obtain the ring $\mathbb{Q}[\sqrt{D}]$.

A commutative ring F is called a *field* if there is always a solution x to the equation $a \cdot x = b$, for all $a \neq 0$ and b in F. (The ring F is assumed not to consist of the single element 0.) Letting $b = a$, we obtain the identity element $x = 1$ of multiplication. Letting $b = 1$, we obtain the multiplicative inverse $x = a^{-1}$ of a.

12.4   Vector spaces

Let F be a field (with elements $\alpha, \beta, \ldots$). A set V (with elements $u, v, w, \ldots$) is a *vector space* over F if any two elements of V can be added, and any element of V can be multiplied by elements of F, in such a way that the following properties are satisfied

I	$u + v$ and αv lie in V.	closure
II	$u + v = v + u$.	commutativity
III	$(u + v) + w = u + (v + w)$.	associativity
IV	There exists $O \in V$ such that $O + u = u + O$.	additive identity
V	$(-1)u \in V$ is such that $u + (-1)u = O$.	inverse of sum
VI	$\alpha(u + v) = \alpha u + \alpha v$, $(\alpha + \beta)u = \alpha u + \beta u$.	distributivity of sum
VII	$(\alpha\beta)v = \alpha(\beta v)$.	distribut. of scalar mult.
VIII	$1v = v$.	identity of scalar mult.

In *VIII* above, 1 is the multiplicative identity of the field.

Bibliography

[1] N. L. Biggs, *Discrete mathematics*, Clarendon Press, Oxford (1994).

[2] H. Davenport, *The higher arithmetic*, Cambridge University Press, Cambridge (1999).

[3] M. Liebeck, *A concise introduction to pure mathematics*, Chapman & Hall/CRC, London (2000).

[4] P. Ribenboim, *The new book of prime number records*, Springer-Verlag, New York (1995).

[5] J. Silverman, *A friendly introduction to number theory*, Prentice-Hall, Englewood Cliffs, New Jersey (1996).

[6] *The New Scientist guide to chaos*, N. Hall, Editor, Penguin Books, London (1991).

Index

$3x + 1$ conjecture, 176

abs, 19, 28, 30, 37, 105, 108
add, *see* sum
arccos, arcsin, arctan, 108
associativity, 8, 9, 44, 187, 191,
    213–215

binomial
    binomial, 153, 164
    coefficient, 78
    theorem, 80

Cassini identity, 85
ceil, 108, 109
commutativity
    commutator, 193
    non-commutative multiplica-
        tion operator, 192
    of addition, 187, 191, 213–
        215
    of logical operators, 47
    of multiplication, 213–215
complex conjugate, 104
computer-assisted proof, 76, 158,
    174, 199
conjugate, 105
constants
    π, 2, 89
    $\sqrt{2}$, *see* square root
    e (Napier's constant), 109, 179
continued fractions, 31
convert, 142
cos, cosh, 108
cot, coth, 108
csc, csch, 108

data type
    exprseq, 62
    float, 97
    fraction, 27
    integer, 27
    list, 64
    posint, 203
    procedure, 51
    range, 62
    set, 45
    string, 66
    symbol, 52
de Moivre's theorem, 108
degree, 128
denom, 27, 28, 37, 54, 93, 109, 178
diff, 2
digits
    Digits, 90
    number of, 37, 38, 109
    of rationals, *see* rational num-
        bers
    pseudo-random, 100
Diophantus of Alexandria, 24
discrete logarithm, 209
distance
    in a metric space, 214
    in a vector space, 188
    in $\mathbb{C}$, 104
    in $\mathbb{Q}$, 27
    in $\mathbb{R}$, 96
    in $\mathbb{Z}$, 7
    triangle inequality, 7, 28, 105
distributivity, 215
ditto variables, *see* variables
divisors
    greatest common, 22, 24

number of, 33
pairs of, 20
prime, 33, 125
proper, 20
twin, 21
do-loops
  do, 163
  for, 164, 165
  in, 169
  loop control variable, 165
  while, 169, 170, 175, 178

error
  absolute, 109
  relative, 109
  syntactical, 8
Euclid, 32
  Euclid's algorithm, 206
eval, 52
evalb, 18, 19, 24, 28, 53, 56, 80,
        121, 122, 134, 150
evalc, 105, 108
evalf, 2, 90, 91, 98, 101, 105,
        108, 110, 121, 178
evalm, 189–194, 196, 197
exp, 108, 109
expand, 2, 34, 81, 125, 130, 133,
        138, 139, 151, 169
expressions
  algebraic, 14
  arithmetical, 8
  logical, 46
  nested, 13, 167
  relational, 18

factorization
  Factor, 210
  factor, 37, 121, 133, 141, 151
  ifactor, 33, 34, 37, 93, 125,
        133, 151, 169
Fibonacci numbers, 84
field, 214
  Cartesian product of, 186
  complex, 103
  finite, 206

rational, 26
real, 96
floor, 108, 109
for, see do-loops
functions
  bijective, 49, 78
  characteristic, 52–55, 60, 69,
        149
  Euler totient, 124, 159
  factorial, 1, 77, 78, 80, 84,
        109
  for divisibility, 23
  gamma, 109
  injective, 49
  piecewise-defined, 177
  surjective, 49
  trapdoor, see trapdoor func-
        tions
  user-defined, 50

GAMMA, 108, 110
gcd, 133

Hamilton-Cayley theorem, 195

I, 105
if, 53–55, 156, 177, 178
ifactor, see factorization
igcd, 23, 37, 62, 63, 68, 123, 133,
        156, 166
ilcm, 23, 37, 72, 133
Im, 105
imaginary unit, 102, 210
in, see do-loops
int, 3
integral domain, 213
intersect, 46
iquo, 23, 30, 37, 54, 63, 93, 109,
        133, 174
  third argument, 38, 94
irem, 23, 24, 28, 37, 48, 53–56,
        70, 73, 93, 133, 174, 182,
        202
  third argument, 38, 94
isqrt, see square root

kernelopts, 10

lcm, 133
lcoeff, 128
least common multiple, 23, 131
Legendre symbol, 208
length, 37, 38, 116
libraries
    linalg, 188
      charpoly, 194, 195
      det, 194
      dotprod, 190
      eigenvalues, 194
      inverse, 194
      norm, 190
      rank, 194
      trace, 194
      transpose, 194
    numtheory, 207
      legendre, 208
      mlog, 209
      msqrt, 208
      order, 208
      phi, 159
      primroot, 209
    standard library, 37
limit, 2
linalg, *see* libraries
log, log10, log[b], 108

map, 56, 120, 121, 179, 196, 207
    optional arguments, 62
max, 108, 109
mediant, *see* rational numbers
member, 47, 175, 176
Mersenne numbers, 35
    largest known prime, 35
    Lucas' test, 35
    mersenne, 36
    primes, 10, 35, 41
metric space, 214
min, 108, 120
minus, 46, 47, 124, 175, 208
mod, 105, 203–210
modular inverse, 205

mul, *see* product
multiplicative order, 208

nops, 45, 120, 122, 124, 174, 175, 180, 181, 210
normal, 138, 141
NULL, 62, 167–169, 174, 178
numer, 27, 28, 37, 54, 93, 109, 178
numtheory, *see* libraries

op, 98, 113–115, 121, 123, 125, 156, 189, 197
operators
    arithmetical, 8
    arrow, 51
    assignment, 14
    logical, 46–48
    non-commutative multiplication, 192
    relational, 18
    selection, 115, 175
    sequence $, 62, 175
    set, 46
option, 182

partition, 202
Pascal's triangle, 79, 164
permutations, 78, 80
Pi, 2, 90, 111
piecewise, 178
plots
    multiple, 72
    of a complex sequence, 106
    of a real function, 111
    of a sequence, 65
    options
      linestyle, 72
      scaling, 107, 124
      title, 66
    plot, 3, 64, 65, 68, 70, 111
pointer, 51, 189
polynomials
    in several indeterminates, 135
    irreducible, 132
    monic, 132

roots of, 137, 210
primes, 32
    gaps between, 65
    isprime, 34, 35, 48, 54, 56,
        61, 120, 122, 150
    ithprime, 36, 37, 61, 65, 66,
        73, 170, 182
    largest known, *see* Mersenne
        numbers
    largest known twins, 41
    nextprime, 36, 37, 180
    prevprime, 36, 37
    primality test, 34
    twin, 41, 50, 122
primitive root, 209
print, 166, 170
proc, 179–182
product
    double, 155
    inert (Product), 153
    numerical (mul), 149–151
    symbolic (product), 152
proof by contradiction, 102, 123
pseudo-random numbers, 100
Pythagorean triples, 24

Quo, 210
quo, 133
quotient of division
    for integers, 20, 22, 38, 67,
        69–71
    for polynomials, 132, 134

rand, 101
rational numbers
    Archimedean property, 28
    digits of, 89, 92–94
    fractional part, 28
    integer part, 28
    mediant, 30, 95, 178
    midpoint, 30
    rational approximation, 99, 178
    reduced form, 27
Re, 105
Rem, 210

rem, 133, 134
remainder of division
    for integers, 22, 38, 48, 67,
        69–71, 73, 203
        divisibility testing, 24
    for polynomials, 132, 134
        divisibility testing, 134
remove, 121
ring, 214
    of Gaussian integers, 215

scalar product, 187
sec, sech, 108
select, 120–123, 181, 210
seq, 61, 62, 65, 68, 74, 80, 94, 97,
    115, 119–121, 123, 134
sequences
    affine, 70
    chaotic, 168
    doubly-infinite, 59
    eventually periodic, 91, 173
    Farey, 73
    fast-growing, 87, 88
    Fibonacci, 84
    linear, 70
    modulation of, 71
    notation, 59
    of digits, 89
    of polynomials, 134
    of sets, 73
    periodic, 67, 68, 71, 102, 107,
        198, 208
    random, 73
    recursive, 82, 94
sets
    cardinality, 40
    difference, 43
    empty, 40
    intersection, 43
    power set, 42
    subset, 42
    union, 43
sign, 108
simplify, 110, 131, 138, 140, 141
sin, sinh, 108

solve, 3
sort, 174
square root
    digits of $\sqrt{2}$, 101
    irrationality of $\sqrt{2}$, 102
    isqrt, 37, 63
    Newton's method, 99
    sqrt, 2, 101, 109, 178
Stirling's formula, 109
substitutions
    sequential, 17
    simultaneous, 17, 83
    subs, 16, 17, 52, 117, 141,
        153
    subsop, 116, 118
sum
    double, 154, 155
    inert (Sum), 153
    numerical (add), 149–151, 156,
        174, 195
    symbolic (sum), 2, 152

tan, tanh, 108
time, 181
trapdoor functions, 34, 209
triangle inequality, *see* distance
triangular numbers, 100

union, 46, 176

value, 154
variables
    ditto, 12, 15, 164
    global, 134
    local, 180
vector space, 215
visible points, 122, 156

whattype, 27, 45, 51, 52, 62, 64,
        97, 113, 118, 151, 189,
        190, 197, 210
while, *see* do-loops
with, 188, 207